Watershed Instrumentation

as it pertains to

Water Quality Analysis, Sedimentation and Hydrology

revised by

David Crutchfeld

edited by

Vladimir Bolshakov

Hydrologic Engineering

Wexford
2008

Table of Contents

Table of Contents (continued)

Table of Contents (continued)

Table of Contents (continued)

Table of Contents (continued)

Table of Contents (continued)

List of Tables

CHAPTER ONE

Introduction

The principles and methods explained in this book are applicable to general watershed analysis. However this was originally written for use in developing countries of the ASEAN (Association of Southeast Asian Nations) region. These are characterized by tropical rainy climates. There generally are no months with a mean temperature less than 18 to 20° C. Annual rainfall normally is abundant, seldom less than 750mm. A tropical rain forest climate, with uniformly high temperatures and and heavy, uniformly distributed rainfall is found in the Philippines. The other countries in this region are located in a monsoon rain forest climate, characterized by a relatively long season of evenly distributed rainfall and a short dry season. Regardless of the amount of rainfall, the ASEAN region often has critical flooding and sedimentation problems. These problems occur, in large part, to the indiscriminate destruction of the forest and other natural resources on upland watersheds.

To effectively plan and implement the integrated conservation and management practices that are needed to achieve a sustained yield of forest, other natural and agricultural resources, specific data sets are required on rainfall, streamflow, sediment, water quality, and the climate of watershed lands. This manual was prepared to help meet these objectives.

This manual will provide a comprehensive overview for watershed planners and managers. It is equally intended to serve the needs of watershed engineers who are responsible for the collections of hydrological and related climatological data sets, and the installation and maintenance of local hydrological and related weather stations or networks of stations. In addition, portions of this manual have been expanded beyond strictly the instrumentation and measurements needs in watershed management to increase its usefulness as a reference for managers, planners, and researchers.

1.1 PURPOSE OF THE MANUAL

The purpose of the manual is to improve both the reliability and comparability of data sets that are collected at hydrological and weather stations in the ASEAN region. To accomplish this, the manual offers standards, procedures, and practices that are designed to minimize the major sources of errors commonly associated with the collections of these data sets. These sources of errors generally are identified as:

- Instrumental errors
- Observational errors
- Sampling errors

The standards, procedures, and practices that are provided to deal with these sources of errors are consistent, to the extent possible, with the well-established practices of the World Meteorological Organization of the United Nations, and other organizations and agencies that are responsible for the collection of hydrological and climatological data sets to quantify the hydrologic cycle on a watershed basis.

1.2 THE HYDROLOGIC CYCLE

In large part, this manual has been organized in terms of the rainfall inputs and streamflow outputs of the hydrologic cycle, and the climate that is associated with this cycle. It may be useful, therefore, to review briefly the hydrological components of this cycle in the introduction to the manual.

In one form or another, water occurs practically everywhere, although the quantity of the water can vary from an almost unlimited supply in the oceans to essentially none in the deserts. Water occurs in the atmosphere as vapor, clouds, and rainfall. On the earth's surface, water is found in streams, in lakes, and in the oceans. Water also is found beneath the earth's surface as groundwater.

2

At any point in time, the largest portion of the water supply is stored in the oceans. A constant circulation of water, however, is always taking place. Evaporation from the oceans' surfaces is continuous. Much of the water that is evaporated then condenses and falls directly on the oceans, although a considerable portion also is carried by the wind to land areas, where the water vapor is precipitated as rainfall.

Some of this rainfall is re-evaporated into the atmosphere before it can reach the earth's surface. Another part of the rainfall is intercepted on the surfaces of forests, buildings, and other objects, and part of this then is re-evaporated directly into the atmosphere. Another portion of the rainfall runs off the ground surface into the streams and, eventually, is returned to the oceans. Still another portion infiltrates the ground surface. For this portion, there are a number of outlets.

Part of the rainfall that infiltrates the ground surface is held by capillarity near the ground surface and is evaporated. Another part of it is used by vegetation in the area and then returned to the atmosphere through the process of transpiration. Still another part enters the groundwater aquifer and then slowly finds its way to the streams, appearing as groundwater flow. A final amount percolates to great depths and then appears after long time intervals, often at distant points, as springs, artesian wells, and geysers.

Of the water that reaches the streams, only a portion flows directly to the oceans. Some part of the water is utilized by people and still another part if consumed and then transpired by vegetation along the edges of the streams. The remainder is evaporated from the surface of the streams and lakes through which the streams flow, or seeps into the ground along the stream channel. This last portion can return later to the same channel downstream.

This general sequence of events, which is represented graphically in terms of a system of water-storage compartments in Figure 1.1, is the hydrologic cycle. The hydrologic cycle provides a basic framework for the practice of watershed management. It is important, therefore, that the most appropriate instrumentation and measurements be used in the quantification of the components of this cycle.

1.3   ORGANIZATION OF MANUAL

The manual has been organized into seven sections, with the first section being this introduction. The second section is concerned with rainfall, the

3

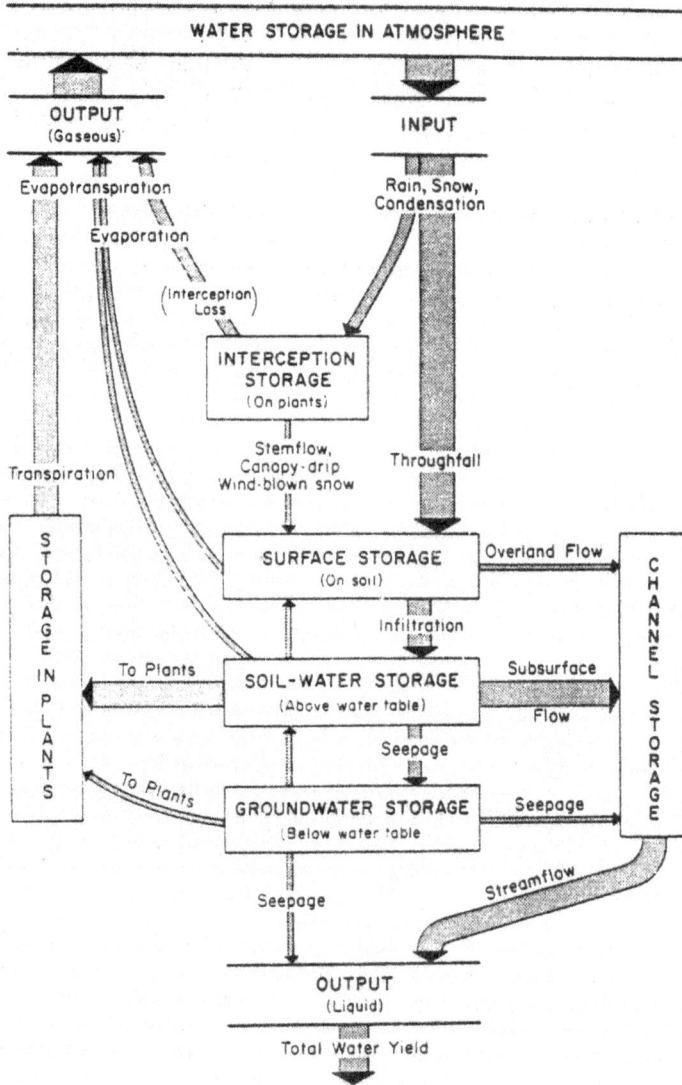

Figure 1.1. The hydrologic cycle consists of a system of water-storage compartments and the solid, liquid, or gaseous flows of water within and between the storage points (Anderson et al. 1976).

4

basic input to the hydrologic cycle. Interception and deposition of the rainfall in forest environments is the topic of section three. Section four of the manual considers streamflow, a basic output of the hydrologic cycle.

Sediment, which are the soil particles transported in streams, and water quality, including the physical characteristics, dissolved chemical constituents, and bacteriological quality of water, are the subjects of sections five and six, respectively.

Climate, specifically considering the collection of data sets for air temperature, relative humidity, evaporation, wind, and solar radiation and radiant energy, is the topic of section seven of the manual.

The organizational structure described above, in the author's judgement, is a suitable way to arrange the subject matter to accomplish the stated objective of the manual. It is recognized, however, that this type of organization, at times, can create difficulties in using the manual in reference to a specific instrument or method of measurement. To facilitate the use of the manual and at the same time to reduce duplication, a cross referencing of the material presented is provided where appropriate.

At the end of the manual, a listing of references is presented to furnish a basis for obtaining further information on the instrumentation and measurements that are used in watershed management. It is beyond the scope of this manual to include the all of details that relate to each of the instruments and measurement procedures described in the following chapters. The listed and other appropriate references should be consulted when this information is required.

CHAPTER TWO

Rainfall

2.1 TYPES OF RAIN GAGES

Rainfall amounts are collected and measured in a gage. Most types of rain gages consist of a right cylinder of known cross-sectional area, with straight sides and a sharp upper edge. The three types of gages used to measure rainfall amounts are a non-recording rain gage, a storage rain gage, and a recording rain gage. Although there are a number of different types of recording rain gages, only the weighing, tipping bucket, and float have gained widespread use.

2.1.1 Non-Recording Rain Gage

The components of a non-recording rain gage are a collector, measuring tube, measuring stick, and overflow cylinder (Figure 2.1). The collector is funnel-shaped at the bottom to channel the rainfall into the measuring tube. It also helps to reduce the evaporation of collected water. The circular area of the measuring tube, minus the cross sectional area of the measuring stick, is one-tenth the area of the collector. Therefore, the depth of the water in this part of the tube magnifies the actual rainfall 10 times, increasing the precision of measurement.

When the rainfall collected exceeds the capacity of the measuring tube, the excess overflows into the cylinder. In these instances, when a reading is made, the observer initially submerges the measuring stick, empties the measuring tube, and then pours this excess into the tube for measurement.

7

The measuring stick, either wood or plastic, is marked such that each 0.01 mm on the stick represents 0.01 mm of rainfall. The waterline on a wood stick is easier to see than on a plastic stick. Since the water is likely to be absorbed by the wood, there may be less chance of erroneous readings due to the waterline being displaced. On the other hand, a plastic stick has

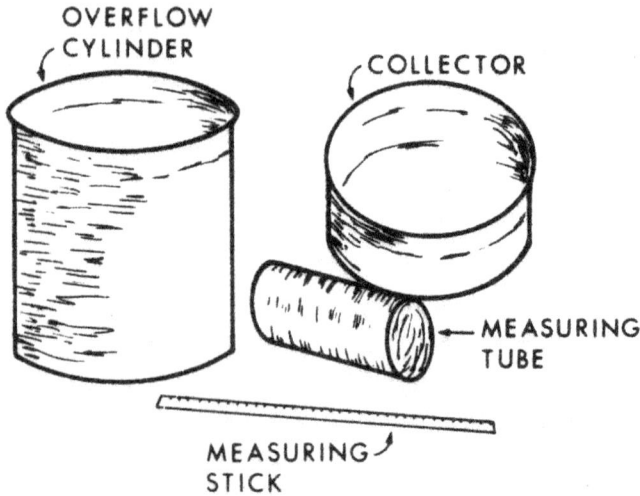

Figure 2.1. Components of a non-recording rain gage (Fischer and Hardy 1972).

advantages over a wood stick, including tne water not creeping up the plastic stick, it is more durable and is easilier to clean.

A non-recording rain gage can be mounted on either a prefabricated steel tripod or a wooden stand. The latter often can be constructed from the shipping container in which the rain gage was obtained.

A non-recording rain gage generally is read at specified time intervals and at the same time each day of measurement. In periods of frequent rainfall events, it may be read at the termination of a rainfall event or, at a minimum, every 24 hours.

2.1.2 Storage Rain Gage

A storage rain gage, also a non-recording gage, has the same diameter at the rim as a standard non-recording rain gage. However, as its name implies, it has a greater storage capacity.

8

A storage rain gage also is read at a specified, but less frequent interval, for example, once a week, once a month, or seasonally. A small amount of oil often is added to a storage rain gage, and sometimes to non-recording rain gages, to suppress evaporation of water that may occur between readings.

2.1.3 Recording Rain Gage

A recording rain gage allows for continuous measurements of rainfall amounts, which then can be used to determine the time duration, intensity, and amount of rainfall for each storm event, as well as the total rainfall amount in any time period. Examples of recording rain gages are the weighing, tipping bucket, and float. All consist of four main components, including a collector, a measuring mechanism, a recording mechanism, and a housing.

a. Weighing rain gage

The operating principle of a weighing rain gage is relatively simple. As rainfall is caught in the collector, it is funnelled into a bucket which is located on the platform of a spring-scale weighing mechanism (Figure 2.2).

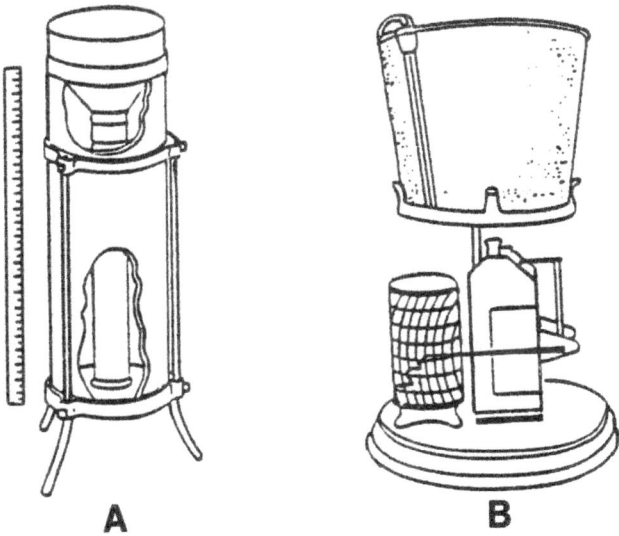

A **B**

Figure 2.2. Illustration of (A) a cut-away of the standard weather service rain gage, and (B) a weighing-type recording gage with its cover removed to show the spring housing, recording pen, and storage bucket (Hewlett 1982).

9

The weight of the collected rain is then transmitted instantaneously through a linkage system to a calibrated pen mechanism on a clock-driven drum. The pen arm traces the weight of the catch of rain onto a chart on the drum, indicating the accumulated rainfall with time. Intensity of rainfall is determined by measuring the incremental changes in rainfall amount within a specified time intervals, typically one hour.

The housing for a weighing rain gage encloses the entire operating mechanism, with the collector serving as the top. A sliding door is located at the bottom of the housing for access to the chart drive and pen arm.

The time scale for the chart is varied by the gear selection for the drive mechanism. Daily, weekly, and monthly time intervals generally are available.

A weighing rain gage to measure both the total rainfall amount and the intensity of the rainfall is generally more accurate than can be obtained by a tipping bucket rain gage, and, therefore, its use is becoming more general.

b. Tipping bucket rain gage

A tipping bucket rain gage is used primarily for remotely measuring rainfall. Again, the operating principle is relatively simple. The caught rainfall is funneled from the collector through a small spout to a tipping bucket mechanism consisting of two small buckets. The bucket beneath the spout fills to capacity, loses its balance, and tips. As the bucket tips, it closes a switch, which transmits an electrical impulse to a remotely located recorder. As the filled bucket tips, the second bucket is positioned under the spout. The collected rainfall from the tipping bucket is emptied into an overflow reservoir.

There are two objections frequently associated with the use of a tipping bucket rain gage. One objection relates to the design of the buckets themselves. If the buckets have been designed to tip at precisely the right instant for a given intensity of rainfall, they will tip either too soon or too late for other rainfall intensities, because of inertia. As a result, both the intensity and the total amount of rainfall recorded will be in error. A second objection to a tipping bucket rain gage is the inconvenient form of the rainfall record obtained when the intensity of the rainfall is high. In these instances, the bucket tips so rapidly that the record becomes difficult, if not impossible to read.

c. Float rain gage

A float rain gage is similar to a weighing rain gage, in that the pen is actuated by a float on the surface of the collected rainfall in the receiver in the same manner as the weighing rain gage. The record of a rainfall event obtained by a float rain gage also is in the form of a mass diagram, as obtained by a weighing rain gage.

2.2   SITE REQUIREMENTS AND INSTALLATION OF A RAIN GAGE

Site requirements for the location of a rain gage is the primary consideration in obtaining accurate measurements of rainfall. An ideal location is one where no eddy currents are likely to occur near the gage. A rain gage location should be far enough away from surrounding trees and other obstructions so that these objects do not affect the collection of rainfall. If a rain gage is located too close to trees and other obstructions, the wind patterns in the vicinity of the gage can cause the rainfall collections in the gage to be considerably different than the rainfall amounts that actually occurred. As a general rule, a clearing defined by a 30 to 45 degree angle from the top of the gage is suggested as the installation site for a rain gage (Figure 2.3).

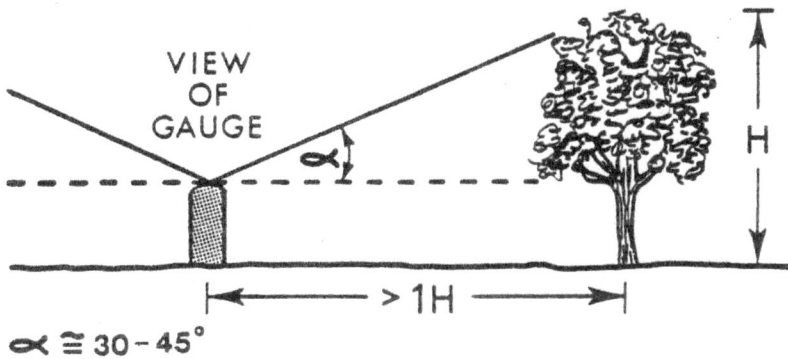

Figure 2.3. Proper siting of a rain gage with respect to the surrounding objects (Brooks et al. 1990).

A rain gage generally should be located on a flat area, with the opening oriented so that the rainfall is measured in terms of the depth that accumulates over a flat surface. As a guideline, the orifice of the gage should be situated approximately 1 m above the ground surface.

Sometimes, a rain gage must be installed in situations where the effects of wind on the rainfall catch cannot be minimized through the site selection, or where a rapid tree growth can affect an opening that was large enough when the rain gage was installed. In such cases, a shielding vane, which reduces eddy effects in areas of high wind speeds, should be used. If a shielding vane is utilized, the top of the vane should be only 1 cm above the receiver funnel.

Detailed instructions for the assembly of a rain gage, regardless of the type, should be adhered to carefully for proper installation. Supports, enclosures, and other accessories also may be necessary to ensure the obtainment of quality records of rainfall.

2.3 RAIN GAGE NETWORKS

Rain gages should be located in a network throughout a watershed so that spatial and elevational differences in rainfall can be measured accurately. The number and distribution of rain gage locations in the network, therefore, are primary considerations. Practical considerations, including accessibility and costs, also limit the type, location, and number of rain gages that can be included in a network.

2.3.1 Number of Rain Gage Locations

Because rainfall amounts can be variable within short distances, it is necessary to install several rain gages in a network to obtain a better estimate of the average rainfall amount for an area. Several rain gages become particularly necessary on forested watersheds, where the variation in rainfall tends to be the greatest.

The number of rain gages needed to measure rainfall generally increases with the size of the watershed to be sampled and the inherent variability of the rainfall events. Sampling requirements can be determined with standard statistical methods.

It often is suggested that random sampling designs be used in estimating the number of rain gages, especially on watersheds where the prevailing storm type is cyclonic. After preliminary measurements of rainfall from several gages for a specified period are available, the number of rain gage locations required to estimate the average rainfall amounts and other parameters for a specified degree of accuracy, assuming that random locations will be selected, can be determined by:

$$n = \frac{(s)^2(t)^2}{(S_x)^2} \qquad (2.1)$$

where n = number of rain gages
 s = standard deviation

 t = statistic for the desired probably level and n-1 degrees of freedom, obtained from a t-distribution table found in reference books on statistical methods
 s_x = standard error of the mean

The number of rain gages required can be reduced, theoretically, by changing them to other random locations frequently or after every rainfall event.

In situations where heavy rainfall frequently occurs as a result of thunderstorms, the minimum number of rain gages generally suggested to sample rainfall adequately are:

Size of watershed area (ha)	Minimum number of rain gage locations
0 to 15	1
15 to 40	2
40 to 80	3
80 to 200	1 per 40 ha
200 to 1,000	1 per 100 ha
1,000 to 2,000	1 per 250 ha
2,000 and over	1 per 750 ha

These minimum number of rain gages are only general estimates. Special considerations, such as the extreme localization of storm patterns, would impose a higher minimum requisite. Importantly, these minimum numbers are not intended to preclude considerations of the needs and conditions of a specific watershed area.

2.3.2   Distribution of Rain Gage Locations

Where the prevailing storm type is cylonic (generally, rainfall of low intensities on a large area), a rather sparse rain gage network can be adequate. In this instance, as mentioned above, a random sampling design can be appropriate. A random sampling design also can be used in situations where the prevailing storms are convectional in nature (with thunderstorms often accompanied by high rainfall intensities), although a fairly dense network of rain gage locations will be required. However, where the prevailing storm type is orographic, modifications in the distribution of rain gage locations may be necessary.

Where the prevailing storms are characterized as orographic, as frequently is the case in rugged mountainous areas, it is necessary to provide a sufficient number of rain gage locations to sample rainfall in the different elevational zones on a watershed. In achieving this sampling requirement, the watershed initially may have to be stratified into the desired elevational zones to be sampled, after which a random sampling design is employed in the strata.

The close proximity of high topographic barriers between the source of the rainfall and the watershed can require a denser rain gage network than where no barriers exist, regardless of the sampling scheme. In this instance, an appropriate compensation in the sampling design generally is necessary.

2.4   DATA COLLECTION AND PROCESSING

2.4.1   Time and Frequency of Data Collection

Rainfall characteristics of interest generally include the total amount in an arbitrarily-defined time period (for example, daily, monthly, seasonally, or annually) or in terms of specific events, the intensity of rainfall, and the distribution of rainfall in time and space. Although rainfall characteristics often are measured routinely at airports and major cities, these characteristics may not be measured on a routine basis in remote areas of watersheds. Therefore, it may be necessary for a watershed manager to decide on the time and frequency of rainfall data collection, a decision that in many cases is arbitrary.

Regardless of the frequency of rainfall data collection, unfortunately, observations often are not made at a specified time of the day, creating a problem in data interpretation. Measurements of a non-recording rain gage made in the late afternoon may be recorded as rainfall for that day, although actually it can represent the rainfall that fell during the preceding 24 hours. Records obtained from a recording rain gage can be interpreted subsequently to represent rainfall characteristics occurring from midnight to midnight, although these readings also could be taken in the morning and recorded as rainfall falling on the preceding day. It can be seen, therefore, that at three adjacent rain gage locations, a 24-hour rainfall event that may have fallen, for example, on September 15, might be recorded at one location as having fallen on September 14 and 15, at another location on September 15 and 16, and at a third location on September 15 only.

Because of this situation, it can be necessary to adjust the rainfall records obtained at some rain gage locations, regardless of how obtained, so that the daily values for all rain gage locations will represent the same 24-hour period.

2.4.2 Sources of Errors in Data Collection

In addition to the problems of adjusting rainfall data to represent the same 24-hour period, two types of errors should be considered in determining rainfall characteristics: instrumental error and sampling error. Instrumental error is related to the accuracy with which a rain gage will catch the "true" rainfall amount at a point. Sampling error is a measure of how well the rain gages in a network on a watershed represent the rainfall characteristics on the entire watershed area. Care in siting a rain gage correctly and properly maintaining can help to minimize instrumental error. Sampling error is minimized by properly designing a rain gage network with an adequate number of rain gage locations, as discussed above.

2.4.3 Data Processing

Processing of rainfall data that are collected from a non-recording rain gage is a straightforward procedure. The measuring stick simply is placed into the measuring tube, with the rainfall collected read directly and then recorded on forms. Processing of rainfall data that are obtained from a recording rain gage and recorded onto charts requires a more detailed procedure, however.

Processing of charts from a recording rain gage should begin as soon as the charts and supporting field notes are received in the office. The processing procedure frequently will be done in three separate operations, often by different individuals. The first step is chart annotation, that is, the transfer of all the field notes and other notes that may be helpful to the proper compilation of the data onto the appropriate charts. Next, the tabulation of the data is undertaken, and finally, the calculation of the depths and intensities of rainfall is made.

a. Chart annotations

Initially, the charts should be numbered and dated to show that the rainfall record is continuous, even though no rainfall occurred during the period of the record. For those charts that cover periods in which no rainfall occurred, it is necessary to only show the chart number and the dates. No other notes are needed, as the primary purpose of these charts is to show a continuity of the records.

For those charts obtained from a recording rain gage equipped with a weekly gear that covers periods receiving rainfall, the following general procedure should be considered:

- Enter the chart number and dates.
- Note the date on the centerline of each day to facilitate the determination of days when rainfall occurred.
- Show the time zone.
- Enter the dates and watch times of chart placement, inspections, and removal.
- Enter notes to show the measured depth of rainfall (not the chart reading) collection at each inspection and removal. If the location also has a standard non-recording rain gage, enter the amount of rainfall measured in this gage.
- Be sure that the chart shows the designation of the rain gage location and the name of the observer, if not recorded before taking the chart to the field.
- Enter notes to show any malfunctioning of the instrument or clock.

For those charts from a recording rain gage equipped with 6-, 12-, or 24-hour gears that cover periods receiving rainfall, the following procedure is followed:

- Enter the chart number and dates.
- Show the time zone.
- Enter the dates and watch times of chart placement and removal.

16

- Show the measured depth of rainfall collected at time of removal. If location also has a standard non-recording rain gage, enter the amount of rainfall measured in this gage.
- If the measured depth of rainfall collected does not agree with the chart line within 1 percent, note the depth corrections to be applied to the chart readings. If after applying the required constant correction for the zero setting the corrected chart reading still does not agree with the measured depth of rainfall collected, distribute the remaining error proportionally.
- Check to determine whether the chart placement and removal mark agree with the watch time. If they do not agree within a couple of minutes, a time correction should be applied. A straight-line variation typically is assumed between chart placement or inspection, and removal.
- Enter the dates and corrected watch times of the beginning and ending for each period of rainfall.
- Be sure that the chart shows the designation of the rain gage location and the name of the observer.
- Enter notes to show any malfunctioning of the instrument or clock.

b. Tabulation of data

Each watershed should be evaluated independently, to determine the detail level needed for accurate rainfall tabulations. For smaller watersheds, it may be feasible and desirable to tabulate time-intervals of 2 minutes. On larger watersheds, where short periods of high intensity rainfall have less effect on a hydrograph, the minimum time-interval might be 5 to 10 minutes. The criteria also might be to tabulate in detail all of the rainfall events in excess of a minimum amount, plus the rainfall events that result in surface runoff.

An individual rainfall event on a small watershed may be considered as one that does not cease for a period of more than 1 hour, after having started. The period of cessation may be increased for larger watershed areas.

The storm records from the charts, corrected in accordance with the above annotations, should be tabulated on forms designed specifically for this task, an example of which is illustrated in Figure 2.4. Entries on these forms should be made in accordance with the following general instructions:

- Fill in all data indicated on the form. Include the designation of the rain gage location, year, and storm numbers. The dates should

TABULATION OF RAINFALL DATA

RAIN GAGE:_____ DATE: _____ TO: _____

TIME:_____ LOCATION: _____ ELEVATION:_____

DATE	TIME	TIME INTERVAL	ACCUMU-LATED DEPTH	DEPTH FOR–		INTENSITY FOR–		REMARKS
				Each time interval	Regular time interval	Each time interval	Regular time interval	
(1)	(2)	(3)	(4)	(5)	(6)	(7)	(8)	(9)
	hr&min	minutes	mm	mm	mm	mm/hr	mm/hr	

Figure 2.4. Example of a tabulation form for rainfall data.

18

cover all of those included in the tabulation, even when no rainfall occurred on many of the days.

- Since the tabulations should indicate a continuity of records, show the date of the first storm of a year and the date of the last storm of the previous year on the first tabulation sheet for the year.
- Mark off the "break points" on the chart. Disregard "minor" changes, especially during periods of low rainfall intensity.
- Tabulate the correct time to the nearest minute of the beginning, break points, specified time-intervals, and end of each storm, in column (2) in the example.
- Tabulate the correct depth of rainfall collected for all of the tabulated times, in column (4). Leave a space after the end of each storm and enter the amount of rainfall actually measured in the collector.
- Enter the name of the person making the tabulations and the date at the bottom of the sheet.

c. **Calculations**

The general procedure to follow in calculating rainfall depths and intensities is outlined below:

- On the tabulation sheet, enter the differences in minutes between successive entries of corrected times, column (2), in column (3).
- Enter the differences in successive entries of corrected depths of rainfall collected , column (4), in column (5).
- In column (6), show the hourly totals for those rain gage locations with rainfall amounts for regular time-intervals.
- Enter in column (7) the rainfall intensity for each time-interval, which is obtained by dividing the values in column (5) by the time-interval value in column (3) and multiplying by 60. (A table of rainfall intensities for given depths and time-intervals can simplify the computation.)
- Fill in the "maximum rainfall depth and intensity" at the bottom of the sheet for those storms that have intensities given in column (7). This task is accomplished by checking column (7) to find the maximum depths for the highest intensities for the various durations.
- Show in column (9), opposite the proper time in column (2), the statement "rain began" or "rain ended." Column (9) also is used to record the daily rainfall amount and any descriptive information.
- Enter the name of the person making the calculations and the date at the bottom of the sheet.

2.4.4 Missing Data

It is not uncommon that one or more rain gages in a network becomes non-functional for a period of time. When this occurs, one approach to estimating the values of the missing data is to utilize existing relationships with other rain gages in the network. For example, if the rainfall amount for a storm is missing for the rain gage at location A, and rainfall measured at location A is correlated with that measured at locations B and C, the storm rainfall at location A can be estimated by:

$$R_A = \frac{1}{2} \left[\frac{N_A}{N_B} R_B + \frac{N_A}{N_C} R_C \right] \qquad (2.2)$$

where R_A = estimated storm rainfall at location A
N_A, N_B, N_C = normal annual rainfall for locations A, B, and C, respectively, where the normal rainfall is the long-term mean annual rainfall, e.g., 10-year, 20-year, or 30-year average
R_B and R_C = storm rainfall for locations B and C

A more generalized equation for missing data is:

$$P_x = \frac{1}{n} \left[\frac{N_x}{N_1} R_1 + \ldots + \frac{N_x}{N_n} R_n \right] \qquad (2.3)$$

where P_x = estimated storm rainfall at location x
N_x, $N_1 \ldots N_n$ = normal annual rainfall for locations x, 1...n
n = number of other rain gages in the network
$R_1 \ldots R_n$ = storm rainfall for locations 1...n

Importantly, equation (2.3) should not be used if there is not a relatively high correlation among the rain gage locations in a network. In addition, this equation is not recommended for use if rainfall from convectional storms is to be estimated.

2.5 ANALYZING RAINFALL DATA

2.5.1 Mean Rainfall on a Watershed

The mean depth of rainfall on a watershed can be estimated by one of three commonly used methods for this purpose, namely, the arithmetic mean, the Thiessen mean, and the isohyetal method (Figure 2.5). As

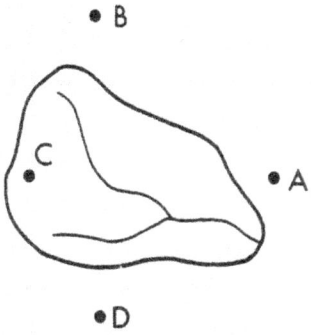

SOURCE DATA:

Daily rainfall measured at each
gage in millimeters

A	B	C	D
40	80	100	60

$$\text{Arithmetic Mean} = \frac{40 + 80 + 100 + 60}{4}$$

$$= 70 \text{ millimeters}$$

THIESSEN POLYGON:

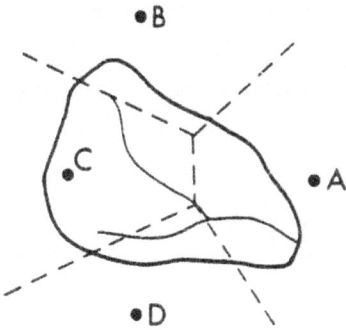

Station	Depth (mm)	Area in Polygon*		Volume (mm)
A	40	x .28	=	11.2
B	80	x .09	=	7.2
C	100	x .49	=	49.0
D	60	x .14	=	8.4
		sum	=	76.0 mm

*As a fraction of total area

ISOHYETAL:

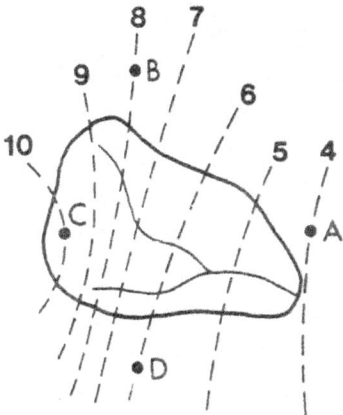

Mean Depth (mm)		Area Between Isohyets*		Volume (mm)
45	x	.12	=	5.4
55	x	.25	=	13.8
65	x	.14	=	9.1
75	x	.13	=	9.8
85	x	.18	=	15.3
95	x	.14	=	13.3
105	x	.04	=	4.2
		sum	=	71.0 mm

Figure 2.5. Methods of estimating the mean rainfall on a watershed (Brooks et al. 1990).

21

discussed below, the first two methods are "mechanical processes" requiring no special skill or judgement on the part of the analyst. However, the accuracy of the results obtained from the third method, which perhaps is the most accurate, depends largely upon the judgement of the person making the computations.

a. Arithmetic mean

An arithmetic mean is the simplest of the three methods for estimating the mean rainfall on a watershed. As the name implies, this value is obtained by dividing the sum of the depths of rainfall recorded at all rain gage locations on the watershed by the number of locations (Figure 2.5). If these locations are distributed uniformly on the watershed and the rainfall varies in a "regular" manner, the results obtained by this method will not differ greatly from those obtained by the other methods. However, if the rain gage locations are irregularly spaced and relatively few in number, and the rainfall varies considerably, more accurate results may be derived by employing one of the other methods.

b. Thiessen mean

When the rain gages in a network are not uniformly distributed on a watershed, the Thiessen mean often can be used to improve the estimates of rainfall depth of the watershed. In the application of this method, adjacent rain gages are connected by straight lines, dividing the watershed into a series of triangles (Figure 2.5). Perpendicular bisectors then are drawn on each of the lines, forming a series of polygons, each of which contains one rain gage. The entire area in a polygon is nearer to the rain gage contained therein than to any other rain gage, and, therefore, it is assumed that the rainfall recorded at the location applies to that area.

The results obtained from a Thiessen mean calculation usually are more accurate than the arithmetic mean when the number of rain gages on a watershed is limited, and when rain gages are located outside of the watershed but can be utilized to represent the rainfall.

The Thiessen mean calculation also assumes a linear variation of rainfall between the rain gages and makes no allowances for orographic influences. Nevertheless, once the area-weighing coefficients are determined for each of the rain gages on a watershed, they become fixed and the Thiessen mean calculation is as simple as the calculation for the arithmetic mean.

c. Isohyetal method

In the isohyetal method, rain gage locations and rainfall amounts recorded are plotted on a map, after which isohyets (contours of equal rainfall amounts) are drawn (Figure 2.5). Measurements of rainfall that are obtained at locations both inside and outside of a watershed can be utilized to estimate rainfall amounts and to draw isohyets. The average depth of rainfall then is determined by computing and dividing by the total area.

Many watershed managers think that the isohyetal method, theoretically, is the most accurate of the three methods for estimating the mean rainfall on a watershed. However, this method also is the most laborious and generally requires the most extensive rainfall data from inside and outside of a watershed to draw accurate.

The accuracy of the isohyetal method depends largely upon the skill of the analyst. An improper application of the method can lead to serious errors. If a linear interpolation between rain gages is used, the results of the isohyetal method essentially will be the same as those obtained with the Thiessen mean calculation.

2.5.2 Double Mass Analysis

If a rain gage has been moved and, therefore, one of the rainfall records is inaccurate, it is desirable to check the accuracy and consistency of the record against that of one or more nearby rain gage locations. This check can be accomplished by plotting the accumulated rainfall measured at the rain gage in question against the average accumulated rainfall measurements obtained from one or more nearby rain gages that are influenced by the same meteorological conditions. The application of this check, referred to as a double mass analysis, can be illustrated with an example.

Assume that the rain gage at location E has been collecting rainfall measurements for 50 years. Originally, location E was a large opening in a forest, but over the years, the surrounding forest has grown to the point where it is suspected that the catch of rainfall in the rain gage is affected by the surrounding trees. Knowing the rainfall patterns in the region, it is recognized that the rain gages at locations H and I are influenced by the same meteorological conditions, although their elevational differences cause the measured rainfall amounts to differ. Fortunately, there was a consistent relationship between the average rainfall measurements at locations H and I with the rainfall measurements at location E. By plotting the accumulated annual rainfall at location E against the average accumulated rainfall at locations H and I, it can be seen that the relationship changed in, for example, 1970 (Figure 2.6). This relationship then can be used to "correct" the more recently obtained rainfall measurements at location E, so that it better represents the "true" catch of rainfall without the interference of the surrounding trees.

2.5.3 Frequency Analysis

It is important to know the probability that a rainfall event of a certain magnitude will occur in a specified period of time to design such structures as small reservoirs, spillways, irrigation networks, and drainage systems. However, as storm systems generally vary from one year to another, it is difficult to predict what the next year, season, or month will bring. Therefore, a watershed manager often relies on the analyses of data on rainfall amounts occurring in specified periods of time. From these analyses, the frequency distributions of past rainfall events are determined, and from these frequency distributions, the probability of a rainfall event of a certain magnitude occurring in a specified time period then can be estimated. This analytical technique, called a frequency analysis, is discussed below.

The purpose of a frequency analysis is to develop a frequency curve, showing the relationship between the magnitude of rainfall events and the associated probabilities of that rainfall event being equalled or exceeded. The probability of a certain rainfall event occurring in a given year can be estimated from a frequency curve. For example, as illustrated in Figure 2.7, the probability of a 24-hour rainfall event of 100 mm or more occurring in a given year is 0.02. Rainfall frequency curves can be developed to evaluate either the maximum annual rainfall events or the minimum rainfall events.

Once a frequency curve has been developed, the probability of equalling or exceeding a certain rainfall amount in a specified period of time can be determined. In general, the probability that a rainfall event with probability p will be equalled or exceeded x times in N years can be determined by:

$$P(x) = \frac{N!}{x!(N-x)!} (p)^x (1-p)^{(N-x)} \qquad (2.4)$$

24

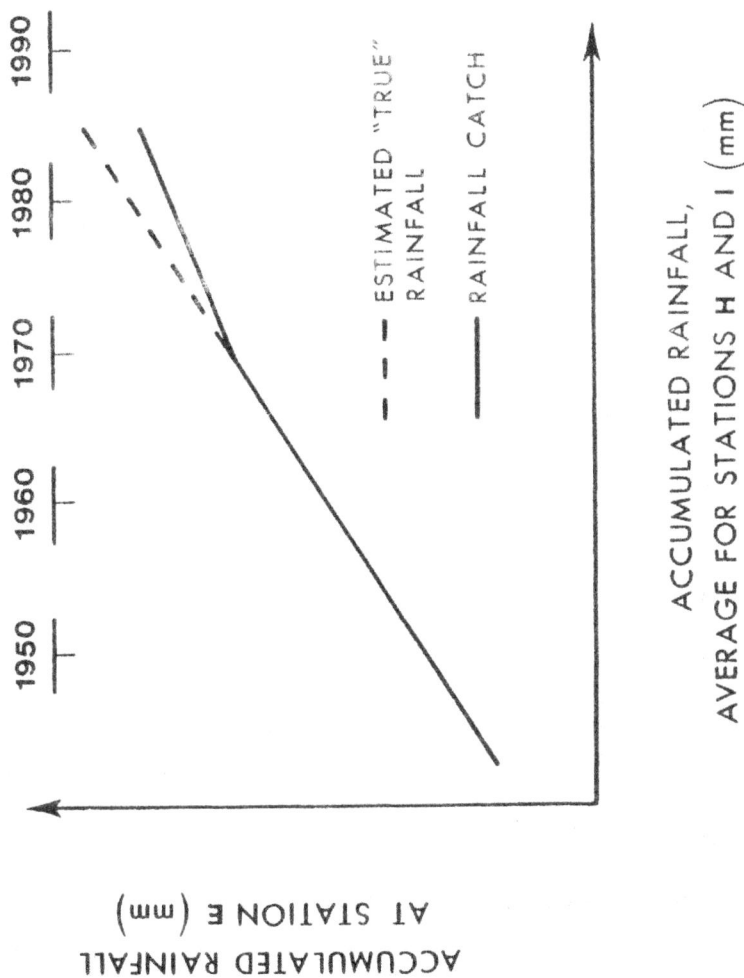

Figure 2.6. Double mass plot of the average annual rainfall at location E against the average annual rainfall at locations H and I (Brooks et al. 1990).

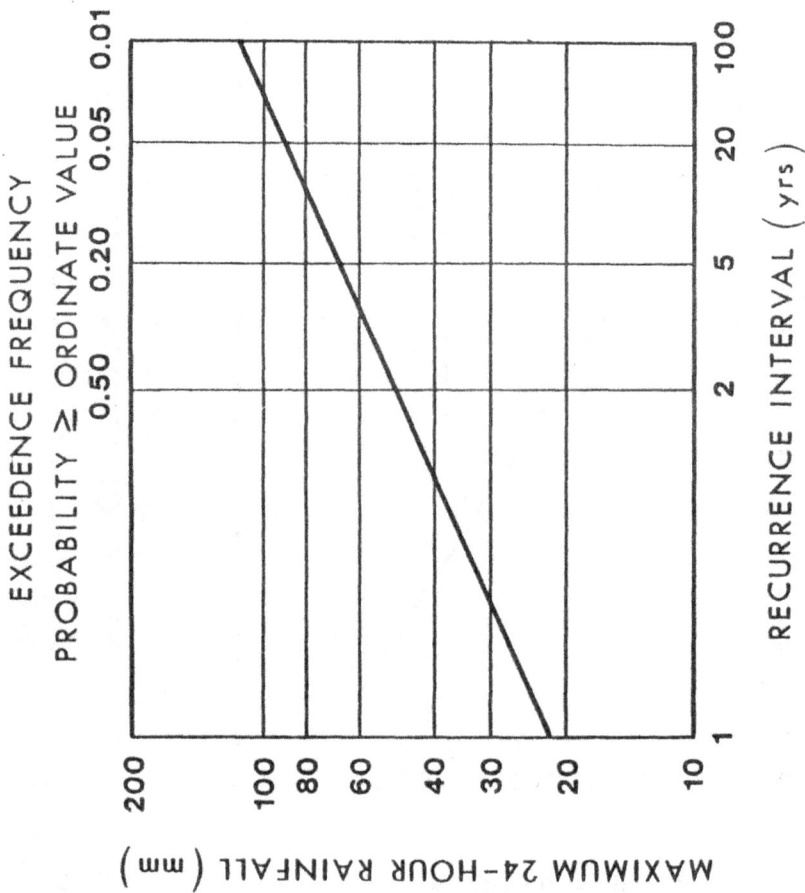

Figure 2.7. A frequency curve of daily rainfall for a single location (Brooks et al. 1990).

If desired, equation (2.4) can be simplified by considering the probability of at least one rainfall event with probability p being equalled or exceeded in N years, that is:

$$P(x=0) = (1-p)^N \qquad (2.5)$$

therefore,
$$P(x) = 1 - (1-p)^N \qquad (2.6)$$

For example, the probability of a 24-hour rainfall event of 100 mm or greater (Figure 2.7) occurring in a 20-year period is determined by:

$$
\begin{aligned}
P(x \geq 100 \text{ mm}) &= 0.025 \\
P(x \geq 100 \text{ mm in 20 years}) &= 1 - (1 - 0.025)^{20} \\
&= 0.40 \text{ or } 40 \text{ percent}
\end{aligned}
$$

2.5.4 Depth-Area-Duration Analysis

A frequency analysis is based upon the rainfall characteristics at a specific location. However, in the design of storage reservoirs, the watershed area must be taken into account. It generally is known, for example, that the larger the watershed area, the lower the average depth of rainfall from a given storm on the entire area (Figure 2.8). In addition, on a given watershed area, the greater the depth of rainfall for a given duration, the lower the expected probability of equalling or exceeding that rainfall amount.

2.6 MAINTENANCE OF RAIN GAGES

A key to obtaining high quality rainfall data is the maintenance of the rain gages in a network. Maintenance requirements for non-recording rain gages are similar and, therefore, can be treated as a group. However, recording rain gages have differing maintenance requirements depending upon the type, make, and model. The manual provided with the instrument should be referred to before maintenance is attempted on recording rain gages.

2.6.1 Non-Recording Rain Gage

The maintenance of a non-recording rain gage is relatively easy. Nevertheless, the simple requirements listed below should be followed to obtain accurate rainfall measurements.

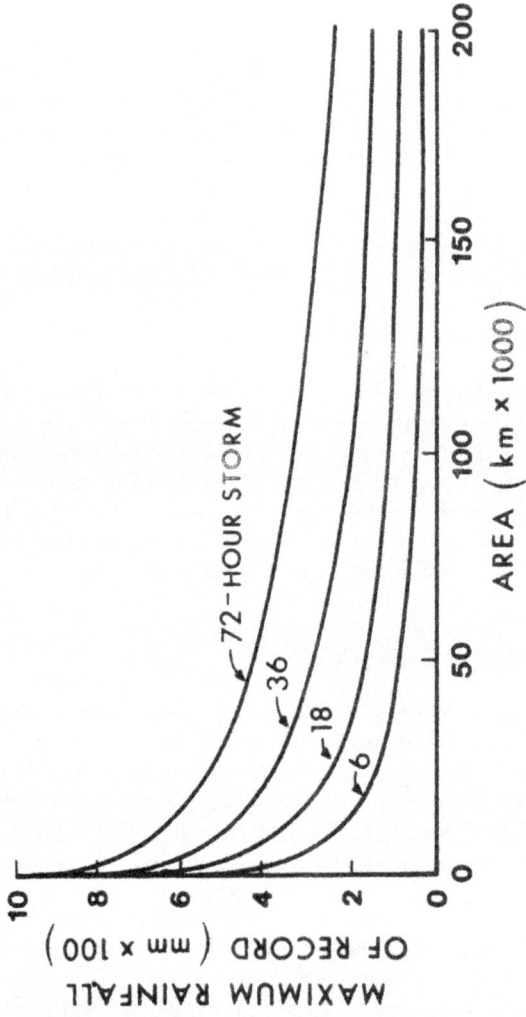

Figure 2.8. Relationship between the maximum rainfall amounts for specified durations and areas (Hewlett 1982).

a. Annual maintenance

Annual maintenance requirements include:

- Carefully check both the measuring tube and the overflow cylinder for irregularities. Repair or replace these components when necessary.
- Check the rim of the collector. It should be perfectly round (except for the small wedge-shaped rain gage), knife-edge sharp, and free of nicks and dents. Repair or replace when necessary.
- Clean the inside of the measuring tube with hot water and a brush.
- Check the condition of the measuring stick. If the markings are faded or if the stick is dirty, clean or replace.
- On plastic instruments that have graduated cylinders, it may be necessary to renew the markings. Marking can be renewed by spreading a small amount of lampblack oil on the surface and then immediately rubbing off the excess with a piece of hard-finish paper.

b. Periodic maintenance

Periodic maintenance requirements are:

- Check the firmness, level, and plumb of the rain gage support. Repair when necessary.
- Keep the top of the rain gage level, checking by laying a carpenter's level across the top of the collector. Adjust when necessary.
- Keep overflow cylinder and measuring tube free of debris. Do not allow debris to accumulate in the funnel. Empty the measuring tube after each measurement.
- Do not handle the graduated part of the measuring stick with hands. Always hold the measuring stick at the upper end.

2.6.2   Recording Rain Gage

Once again, the specific maintenance requirements will vary with the type, make, and model of the recording rain gage. The instrument manual provided by the manufacturer should be consulted for detailed instructions on maintenance and calibration.

a. Weighing rain gage

Maintenance requirements of the weighing rain gage pen, pen arm assembly, chart drive assembly, and clock should follow the specifications of the manufacturer. The following general maintenance items should be performed every 6 months. Again, refer to the instrument manual for detailed instructions:

- Remove the collector and outer case. Clean thoroughly all moving parts, using cleaning solvent applied with a soft brush. Do not use solvents that attack painted surfaces.
- Check the linkage system, spring, and other moving parts for evidence of binding or excessive friction.
- Lubricate the bearing of all moving parts (except the chart drive assembly) sparingly, with a light, non-gumming instrument oil.
- Clean the bucket inside and out to remove accumulated debris.
- Check the level of the fluid in the dash pot. Add the necessary fluid to bring the level to within 0.75 cm of the top of the dash pot. (Dash pot fluid is available from the manufacturer.)
- Check the weighing mechanism for accuracy. (Rain gage calibration weights are available from the manufacturer.)

If the accuracy of a weighing rain gage is questioned, do the following before attempting re-calibration:

- Check the chart installation. The chart must be seated firmly against the flange along the lower end of the chart drum.
- Check the chart drum. It must be seated properly on the spindle or arbor. The external gears must be meshed.
- Check that there are two spacing washers between the base of the rain gage and the large stationary gear at the base of the spindle or arbor.
- Check the mechanical condition of the rain gage, especially looking for points of friction in the linkage.

After checking the above items, proceed to a check of the present calibration status, as follows:

- Place the bucket on the weighing platform.
- Zero the pen arm using the thumb nut. The pen should be on the zero line of the chart.
- Add to the calibration weights the equivalent of a known amount of rainfall, and note the chart value indicated. If the chart values

observed are incorrect for the weight added, the rain gage should be calibrated.
- If there is insufficient or non-uniform motion of the pen over the initial readings on the chart, but correct and uniform motion of the pen is obtained thereafter, the spring probably needs to be replaced.
- Further calibration should not be attempted without consulting the instrument manual provided by the manufacturer.

b. Tipping bucket rain gage

Annual maintenance requirements are:

- Check the rim of the collector. It should be perfectly round and free ot irregularities. Repair when necessary.
- Remove the storage container and then clean and check for leaks.
- Remove the collector and clean all moving parts with a soft brush and cleaning solvent. Do not use solvents that attack painted surfaces.
- Check all parts for wear. Replace when necessary.
- Check the tipping bucket action. Eliminate any binding that might occur.
- Lubricate sparingly the pivots of the bucket and the V-bearing in the support bracket. Use a light, non-gumming instrument oil.
- Do not adjust the position of the calibration screw located in the supporting bracket below the cups, unless the calibration instructions are available.

Periodic maintenance requirements include:

- Keep the collector free of debris.
- Wipe out the tipping bucket with a clean cloth weekly.
- Wipe the bucket pivots and support bracket V-bearing with an oily cloth weekly.
- Check the recorded rainfall with a measured amount of water in the storage container.

c. Float rain gage

The maintenance requirements for a float rain gage generally are similar to those for a weighing rain gage.

2.7 SPECIAL CONSIDERATIONS FOR THE ASEAN REGION

2.7.1 Selection of Type of Rain Gage

An initial decision to be made in the selection of the type of rain gage to use is whether a non-recording or recording rain gage should be installed. When measurements of rainfall amounts are necessary at only periodic intervals, for example, once a week, or once a month, a non-recording rain gage will suffice. When measurements of frequent rainfall events are required, as they often can be in the ASEAN region, a non-recording rain gage must be visited at more frequent intervals, however.

When measurements of rainfall intensities, durations, and amounts are necessary, a recording rain gage should be considered. A recording rain gage, which allows for a continuous measurement of rainfall, is often more costly to purchase, install, operate, and maintain than is a non-recording rain gage. A long-term, committed investment, therefore, must be guaranteed. Nevertheless, where measurements of rainfall intensities, durations, and amounts are necessary to characterize a watershed condition, an investment in a recording rain gage can be justified.

Examples of recording rain gages, as mentioned above, are the weighing rain gage, the tipping bucket rain gage, the float rain gage, and others that were developed for special purposes. For many years, the tipping bucket rain gage commonly has been used throughout the world, although the weighing rain gage also is widely used. In either case and in general, the selection of a recording rain gage is made in situations where the benefits from obtaining measurements of rainfall intensities, durations, and amounts are greater than the relatively high costs that are incurred in making this selection.

Regardless of the type of rain gage selected, it is important that the location at which the rain gage will be installed meets the site requirements. In addition to the location being a relatively flat area not affected by surrounding obstructions, the location must be secured from trespass and easily accessible to the observers. When these conditions cannot be met, an investment in a recording rain gage may be inappropriate and, alternatively, a non-recording rain gage may be selected for installation.

2.7.2 Selection of the Observers

A key to the effective management of watersheds is knowledge of the precipitation inputs onto the watershed. Depending upon the region of interest, these inputs can be rain, snow, sleet, hail, or different combinations thereof. In the ASEAN region, the basic precipitation input is rainfall. It is an absolute necessity, therefore, that this rainfall input be measured accurately, reliably, and completely as possible. To do so requires the selection of observers who both recognize the importance of their tasks and have the necessary training to execute their tasks as required.

Observers should be made to recognize that their work is critical to the adoption of effective watershed management practices. In addition, they should learn to appreciate the place of their work in the general framework of the watershed management practices. A minimum prerequisite to the selection of an observer is an ability to obtain the field records in the manner required, once the appropriate training has been completed. Importantly, the observers also must be reliable people, willing to make the scheduled observations regardless of the weather conditions or the remoteness of the location of a rain gage. To the extent possible, the observers should be well-compensated financially for their efforts.

Inhabitants of watersheds in the ASEAN region frequently are dispersed. Employing the local people as observers, therefore, often can reduce the travel between the observers and the rain gages and, importantly, solve the problems of theft and vandalism.

a. Field observations

Field observations are made to record accurately and as completely as possible what has happened in terms of rainfall events, to take notes on what might be helpful in the interpretation of the data, and to promote the proper operation of the rain gage in obtaining complete records of what is to come.

Notes on the operation of the rain gage are vital to the subsequent tabulation of the data. Notes on the attending conditions are important to the analysis and subsequent interpretation of the data. Many of these extenuating conditions are not recorded through the operation of the rain gage and, hence, must be noted by the observer. Rainfall records from a gage that did not function properly often provide usable information if adequate notes are taken on the nature and cause of failure of the rain gage to operate properly. Observers should be made to feel comfortable about writing their observations and opinions, so that the information not recorded is lost forever.

b. Office work

In some instances, the observers also will be responsible for the reduction of the data or, in other cases, the analysis of the data in the office. Therefore, it may be important for the observers to possess the analytical skills necessary to undertake these assignments, once the required training has been completed.

Observers that are charged with both field observations and office work often make less errors in rainfall data collection, reduction, and analysis than when these tasks are charged to different individuals. Furthermore, watershed management programs in which the observers have all of the full responsibilities for the collection, reduction, and analysis of rainfall data often are more cost-effective than when these watershed management programs are managed otherwise.

2.7.3 Instrument Maintenance Considerations

Maintenance of a non-recording rain gage in the ASEAN region should pose no special problems other than those generally encountered elsewhere. However, because of the high rainfall amounts, the frequency of major rainfall events, and the high relative humidities in the region, the maintenance of a recording rain gage, regardless of the type, make, or model, can call for special considerations.

The general maintenance of a weighing rain gage, usually scheduled every 6 months, may have to be performed at a more frequent interval. In prolonged periods of intensive rainfalls, parts or all of the maintenance requirements should be scheduled more often to ensure the proper operation of the rain gage. Similarly, the annual and periodic maintenance requirements of a tipping bucket rain gage also may have to be scheduled at more frequent intervals. It is important, therefore, that a monitoring scheme to determine when the maintenance of a recording rain gage is required be incorporated into the rainfall data collection schedule.

Whenever the maintenance of a rain gage is required, personnel that are trained in the proper maintenance procedures and, when needed, spare parts and components must be available.

The successful maintenance of rain gage networks on many watersheds in the ASEAN region with illegal occupants, to some extent, would depend upon a good training program. Instrument theft and vandalism from locations that are close to a community of watershed settlers might be prevented by educating these people on the importance of the instruments to them.

CHAPTER THREE

Interception and Deposition of Rainfall in a Forest Environment

The influence of forests on rainfall involves two separate considerations. First, there are the effects of forests on the condensation of water vapor and, therefore, on the rainfall in the air above, in the forests, and for a distance beyond the borders of the forests. Second, there are the effects of forests on the interception and, ultimately, the deposition of rainfall on the soil surface. The first consideration, the effect of forests on rainfall, has been a topic of controversy for a long time. Some have argued that forests augment rainfall and, in doing so, make a climate more humid. Others believe that forests have little appreciable effect on the water contents in the air masses that are the sources of rainfall. Adequate proof generally is lacking in both instances.

The second consideration, the effects of forests on the interception of rainfall, and on the amount and deposition of rainfall that reaches the soil surface, is the topic of this section of the manual. In the forests of the ASEAN region, more than 30 percent of the annual rainfall amount can be lost through interception. Not all the rainfall that is caught by the forest canopy is lost to the atmosphere, however. Much of the water can drip off the foliage or flow down the stems of the trees, ultimately reaching the soil surface. Not all the rainfall that penetrates the forest canopy becomes available to either replenish the soil water deficits or becomes runoff. The forest floor litter can store relatively large quantities of water which, in time, evaporates. It is important, therefore, for a watershed manager to measure the components of interception and, in doing so, to estimate the *net rainfall* amount in the hydrologic cycle. Net rainfall is that available to replenish the soil water deficits or become runoff.

3.1 COMPONENTS OF THE INTERCEPTION PROCESS

As previously mentioned, not all of the rainfall onto forest canopies reaches the soil surface. A part of the rainfall is caught by the crowns of the trees and by other surfaces, and is evaporated back into the atmosphere. The amount that is subtracted from the rainfall amount in this way is known as *interception*.

Little rainfall reaches the soil surface directly in dense forests. The rainfall initially is stored on the leaves, branches, and stems of the trees and other vegetation. Once these surfaces become wetted, additional rainfall will displace the water on the lower edges of the leaves, twigs, and other components. The displaced water falls to the next lower level of vegetation or to the forest floor litter, and eventually to the mineral soil. The amount of water that is stored on the wetted surface of the forest cover, called *interception storage*, is dependent largely upon the form, density, and texture of the leaves, twigs, and other surfaces.

Water moves through a forest canopy by two mechanism, *throughfall* and *stemflow*. Throughfall penetrates the forest canopy directly through the spaces between the leaves or by dripping from the leaves, twigs, and branches. Stemflow reaches the soil surface by flowing down the stems of the trees. These two processes deliver water to the soil surface, where the water must penetrate the forest floor litter layer before entering the mineral soil. Some of the water that reaches the forest floor is stored in the litter, never entering the soil.

During and immediately after a rainfall event, a portion of the water that has been stored in the forest canopy, on the tree stems, and in the forest floor litter is returned to the atmosphere by evaporation. Even during intense rainfall events, when the atmosphere is quite humid, the evaporative loss can be relatively large, because this evaporative loss occurs from a large area of leaf surface.

The components of the interception process are illustrated in Figure 3.1. Interception by a forest canopy can be summarized in the following equation:

$$I_c = P_g - T_h - S_f \qquad (3.1)$$

where I_c = forest canopy interception (mm)

P_g = gross rainfall (mm)

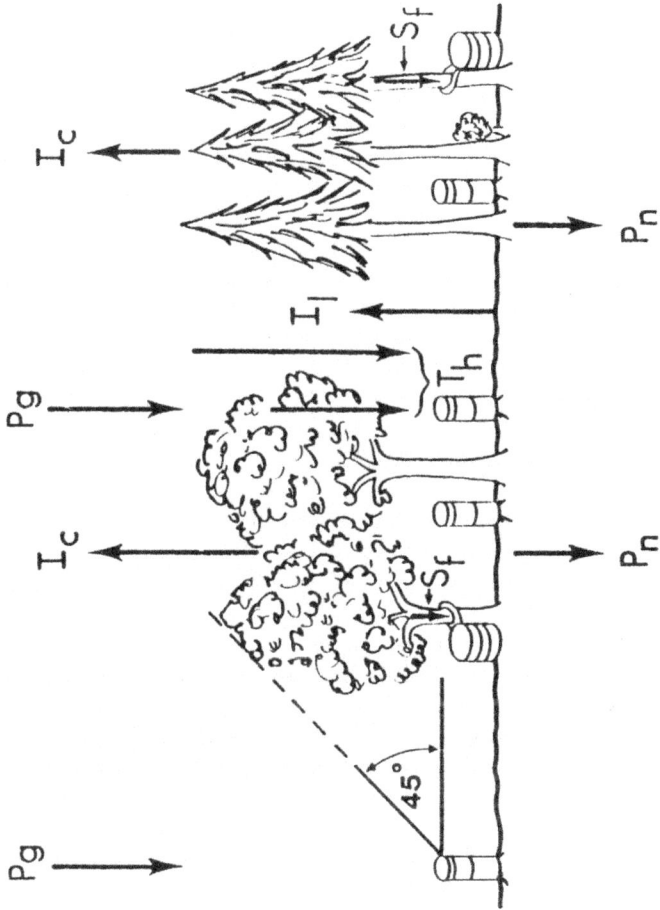

Figure 3.1. Components of interception (Hewlett 1982).

39

T_h = throughfall (mm)

S_f = stemflow (mm)

Again, referring to Figure 3.1 and equation (3.1), the deposition of rainfall on the soil surface can be described in terms of the total interception (I),

$$I = I_c + I_l \qquad (3.2)$$

where I_l = litter interception

the amount of rainfall reaching the soil surface (P_s),

$$P_s = T_h + S_f \qquad (3.3)$$

the amount of canopy interception, considering both overstory and understory vegetative layers in a forest community (I_c),

$$I_c = P_g - T_h - S_f \qquad (3.4)$$

and finally, net rainfall (P_n),

$$P_n = T_h + S_f - I_l \qquad (3.5)$$

Partitioning off specific amount of rainfall into the pathways described in equations (3.2, 3.3, 3.4, and 3.5) depends largely upon the forest cover characteristics, including leaf surface area, shape of the forest canopy, and roughness of the tree bark.

3.2 MEASUREMENTS OF THE COMPONENTS OF INTERCEPTION

The measurements of the components of interception, in most instances, involve the direct measurement of gross rainfall, throughfall, and stemflow, and observations of the changes in the moisture content of the forest floor litter.

3.2.1 Gross Rainfall

Gross rainfall usually is measured in a non-recording rain gage that is located either above the forest canopy, or in a large clearing in the forest or in an adjacent field. A non-recording rain gage located above the forest canopy is mounted on a tower that extends above the heights of the tallest trees in the forest, away from any eddy currents. When the non-recording rain gage is located in a large opening or field, the instrument must be situated far enough away from obstructions so that these objects do not affect the collection of rainfall.

A description and the general operating instructions for a non-recording rain gage are presented in section 2.1.1 of this manual.

Gross rainfall measurements typically are collected either as part of an investigation of the interception processes on a watershed or in a monitoring program. In both cases, the non-recording rain gage, regardless of its location, usually is read at specified time intervals. Processing of the data sets obtained is a straightforward procedure, as described in section 2.4.3 of this manual.

3.2.2 Throughfall

Throughfall usually is measured by placing a pre-determined number of non-recording rain gages in a random pattern underneath a forest canopy in sample plots. Sample plots sizes for the measurement of throughfall commonly vary between 1 and 3 ha. The inherent heterogeneity of the forest canopy affects the amount, intensity, and spatial distribution of the throughfall component and, therefore, the number of non-recording rain gages needed and the size of the plot.

Depending upon the statistical analytical techniques employed, the non-recording rain gages that are used to measure throughfall are moved periodically within and among the plots to sample a range of tree canopy surface area and forest cover descriptors. These descriptors, in turn, can be included in throughfall predictive equations for wider applications.

Troughs also have been used to measure throughfall. The dimensions, design, and placement of the troughs in a forest environment are based upon the anticipated rainfall intensities, and the composition, structure, and density of the forest overstory to be measured. Unfortunately, the spatial variations in this component of the interception process is difficult to estimate with the measurements obtained with troughs.

Throughfall data collections generally are made to coincide with the measurements of gross rainfall, along with the measurements of stemflow. By doing so, "event-based" measurements of the components of interception are available for analysis.

3.2.3 Stemflow

Stemflow is measured directly by placing metal "collars" in notches in the bark around the stems of sample trees, and then collecting and measuring the flow from these collars, as shown in Figure 3.1.

The magnitude of the stemflow component varies, to a large degree, with the species, age, and size of the trees sampled, all of which are related to the roughness of the bark. Tree species with rough bark generally retain more water and exhibit less stemflow than those with smooth bark. As some species of trees become older, the bark pattern becomes increasingly rough. Consequently, sample trees of differing ages should be selected for stemflow measurement.

To place stemflow measurements on an area-basis, sample plots are established. The size and number of the sample plots, and the number of sample trees on a sample plot, are determined from the variations in the forest overstory conditions and the desired accuracy in measurement.

3.2.4 Moisture Content of the Forest Floor Litter

Litter interception and changes in the moisture content of a forest floor litter are more difficult to measure. By repeatedly weighing samples of the litter, however, it often is possible to determine the minimum and maximum values of moisture content and the rate of drying. By weighing samples of the litter during rainfall events, the rate of wetting can be determined. With these measurements, an accounting procedure then can be used to calculate the changes in the "opportunities" to store moisture in the forest floor litter between rainfall events and the rate at which moisture is stored in the litter during the rainfall event.

3.3 ESTIMATION OF INTERCEPTION

3.3.1 Interception Equations

Rainfall events frequently are separated by intervals of varying lengths. Upon the cessation of a rainfall event, evaporation begins to deplete the interception storage. This process continues until all of the available water has been returned to the atmosphere in the form of vapor or until another rainfall event occurs. The total interception during different rainfall events, therefore, is neither constant in quantity or in the percentage of the total rainfall amount.

For a specific watershed area and condition of the forest overstory, an "interception equation" to describe the total interception is of the general form:

$$I = a + b(T) \tag{3.6}$$

where a = interception storage capacity (mm)
b = evaporation rate (mm/hr)
T = duration of rainfall event (hr)

Equation (3.6) is applicable only to rainfall events exceeding the interception storage capacity. As indicated by the mathematical form of equation (3.6), the total amount of interception increases with the duration of the rainfall event. Since the depth of rainfall also increases with the duration, there generally is a correlation between the interception losses resulting from a high intensity, large rainfall event and losses from a low intensity, small rainfall event. However, the percentage of the total rainfall amount that is lost to interception decreases as the amount of rainfall increases.

In addition to the general model presented by equation (3.6), other mathematical models also can be used to empirically quantify the interception process. The coefficients for these mathematical models are determined through a regression analysis, the concepts of which are described in section 7.7.1 of this manual.

3.3.2 Interception Percentage Values

Interception percentage values, which are values that represent the percent of the gross rainfall intercepted by the forest canopy, can be used to estimate the total interception amount from knowledge of the rainfall amount. Interception percentage values have been reported for several forest types in the ASEAN region.

Interception of rainfall in the forests of Thailand ranges from about 65 percent of the rainfall amount in natural teak forests, to approximately 10 percent of the rainfall in the hill-evergreen forests, to less than 5 percent in the dry-evergreen forests. In the lowland rainforests of Peninsular Malaysia, the interception of rainfall is 20 to 25 percent of the rainfall amount. On the average, the interception of rainfall in the humid tropical forests of the Philippines generally is between 60 and 70 percent of the gross rainfall total.

43

Although it is possible to estimate the total amount of interception from the interception percentage values, the ultimate deposition of the rainfall on the soil surface still is unknown.

CHAPTER FOUR

Streamflow

The general term *runoff* refers to the processes and pathways through which *excess water* becomes streamflow. Excess water is that part of the total rainfall amount which runs off the land surface, and that which drains from the soil and is not consumed in evapotranspiration. The collective pathways through which excess water flows on and through a watershed determines the shape and relative magnitude of a *streamflow hydrograph*, which is a graphical representation of *streamflow discharge*, or rate of water movement, in relation to time (Figure 4.1).

4.1 BASIC TERMS AND DEFINITIONS

The most direct and immediate pathway from rainfall to streamflow is the rain that falls directly into the stream channel. This pathway is called *channel interception*, represented as A in Figure 4.1. *Surface runoff*, also referred to as *overland water flow*, is that pathway in which excess water flows over the soil surface, shown as B in Figure 4.1. Surface runoff, which originates from impervious areas or areas where the rate of rainfall exceeds the infiltration rate of the soil, represents a relatively quick response to a rainfall input, second only to channel interception. *Sub-surface flow*, also termed *interflow*, is that pathway in which excess water infiltrates into the soil, but then arrives at a stream channel in a short enough period of time to be considered part of the storm hydrograph, illustrated as C in Figure 4.1. Sub-surface flow generally is considered to be the major pathway in most forested watersheds.

RAINFALL

A - CHANNEL INTERCEPTION
B - SURFACE RUNOFF OR OVERLAND FLOW
C - SUBSURFACE FLOW OR INTERFLOW
D - GROUNDWATER OR BASEFLOW
Q - STREAMFLOW DISCHARGE

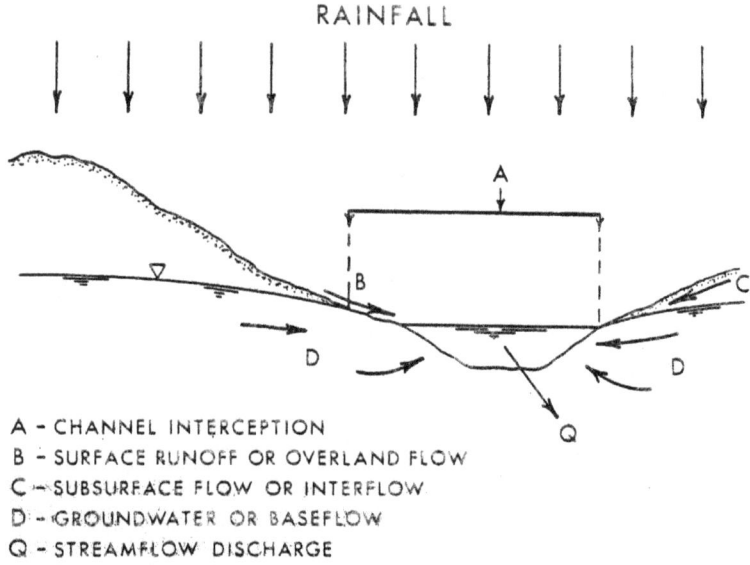

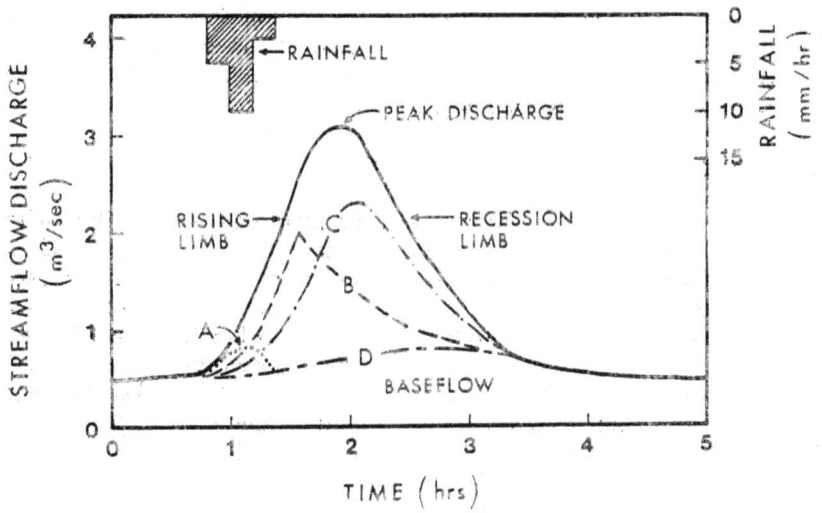

Figure 4.1. Relationship between pathways of water flow from a watershed and the resultant streamflow hydrograph (Brooks et al. 1990).

The sum of channel interception, surface flow, and sub-surface flow is *stormflow*, also called *direct runoff*. Stormflow is that component of a hydrograph that is analyzed when studying the effects of watershed management practices on subsequent water regimes. It often is referred to as the *storm hydrograph*.

In a perennial stream, *groundwater flow*, also referred to as *baseflow*, is a component of a hydrograph, as shown by pathway D in Figure 4.1. Groundwater flow does not contribute directly to the storm hydrograph of an ephemeral stream, however.

In most studies of a streamflow hydrograph, there generally is no attempt made to separate and evaluate the various pathways of flow. Instead, the streamflow response is evaluated through a separation of the stormflow component from the groundwater flow, when the latter is present. Because the hydrograph represents collectively the integrated response to a rainfall event, the separation of a hydrograph in terms of time response, rather than flow pathways, is more useful for analysis.

4.1.1 Units of Measurement

The units of measurement in which the quantities of streamflow discharge are expressed are always volume of water per unit of time. Regardless of the unit of measurement, an integration of the units of streamflow discharge through time gives expressions of the total volume of water flowing from a watershed for the time period integrated.

Common units of streamflow discharge in SI units are:

- Cubic meters per second (m^3/s).
- Cubic meters per second per square kilometer ($m^3/s/km^2$).
- Liters per day ($1/d$).
- Millimeters (or centimeters) of depth on a watershed area per day, month, year, or season.

The corresponding units of streamflow discharge in English units are:

- Cubic feet per second (ft^3/s).
- Cubic feet per second per square mile ($ft^3/s/mi^2$).
- Gallons per day (g/day).
- Inches of depth on a watershed area per day, month, year, or season.

Conversion factors from English to SI units of streamflow discharge are:

$$ft^3/s = 0.0283 \ m^3/s$$
$$ft^3/s/mi^2 = 0.0109 \ m^3/s/km^2$$
$$g = 3.78 \ l$$

4.1.2 A Hydrograph and Its Components

As previously mentioned, a graph of the streamflow discharge against time is called a *streamflow hydrograph*. The basic components of a streamflow hydrograph are illustrated in Figure 4.2 and briefly described below.

The initiation of stormflow is that point at which a hydrograph separates into stormflow and baseflow, and the cessation of stormflow is that point at which the stormflow component returns to the baseflow level. The relative magnitude of the stormflow component of a hydrograph is determined by a number of factors, some of which are fixed and some of which vary with time on a watershed.

Fixed watershed characteristics that influence the stormflow amounts include the size, shape, watershed and channel slopes, the drainage density and network, and the presence of bodies of water within the watershed. Factors affecting stormflow that vary with time are the intensity, duration, amount, and distribution of the rainfall event, the antecedent conditions of the watershed, the type and extent of vegetative cover, and the occurrence of roads, drainage systems, reservoirs, and alterations of the stream channel.

The *time to peak* is the time between the center of mass for a rainfall event and the peak streamflow rate. The rising limb of a hydrograph generally is thought to reflect the storm characteristics of the rainfall event, that is, the intensity, duration, and amount of rainfall, while the recession limb often is indicative of the storage characteristics of a watershed.

4.2 MEASUREMENT OF STREAMFLOW DISCHARGE

Streamflow discharge data are important to a watershed manager in planning for flood control, estimating the dependability of water supplies, or designing reservoir storage. In measuring the streamflow discharge, the initial step is to measure the height of water, referred to as the *stage*, at some point along a stream with a staff gage or a continuous water-level recorder. The stage is referenced to an arbitrary datum, such as a place on

48

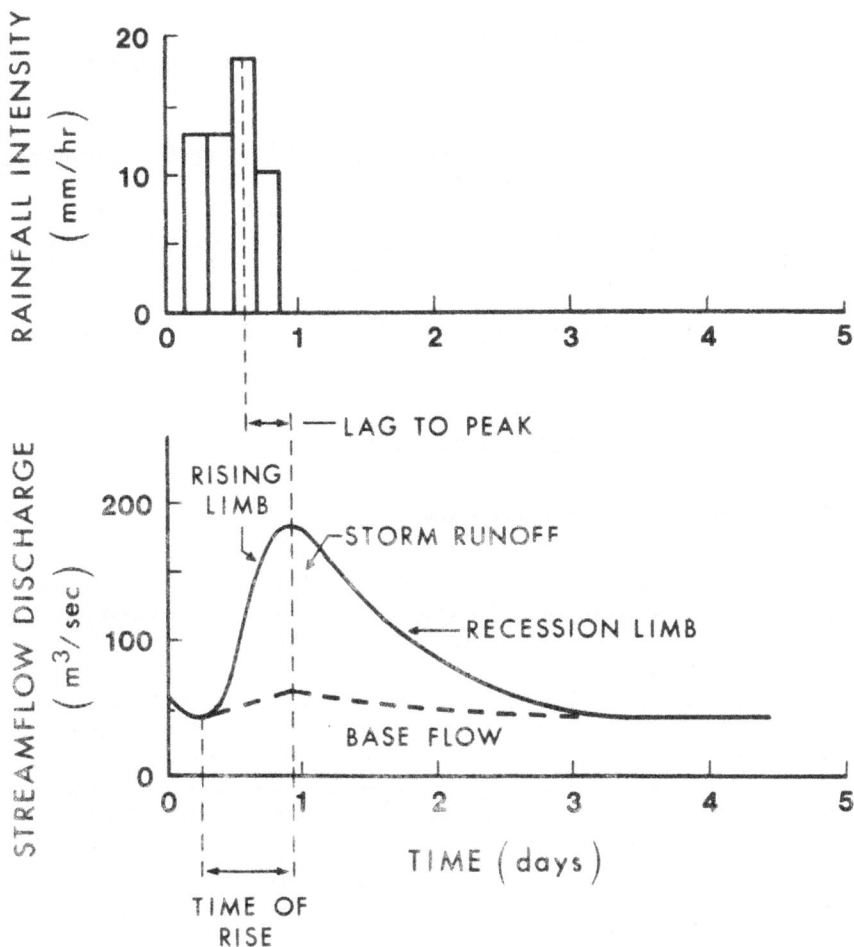

Figure 4.2. Basic components of a streamflow hydrograph (Dunne and Leopold 1978).

the stream bed or on the surface of a precalibrated structure installed for streamflow measurements.

The stage, the gage height, and the level of the water surface all are considered to have the same meaning. However, for the most part, the term *stage* will be used in this manual to indicate the height of water. The stage can be measured with a staff gage or a continuous water-level recorder, of which there are several types.

4.2.1 Staff Gage

A staff gage is a graduated staff used for the visual observation of water levels. A staff gage is used for periodic observations on streams where continuous water-level recorders are either not needed or not feasible. In addition, a staff gage often is used in conjunction with a continuous water-level recorder to provide a check on the performance of the recorder.

A staff gage equipped and protected to retain a high-water line is referred to as a crest gage.

A staff or crest gage requires a minimum of protection, other than that provided through the selection of the control section.

4.2.2 Continuous Water-Level Recorder

A variety of continuous water-level recorders, one of which is shown in ʹFigure 4.3, are available from the manufacturers of weather instruments. Many of these recorders operate on a reversing mechanism for stage that permits the recording of an unlimited range in stage on a scale that can be read accurately. Both float-operated and pressure transmitters generally have been found to be capable of producing reliable water-level measurements, although a float-operated water-level transmitter often can be more cost-effective. An essential requirement of all continuous water-level recorders, regardless of the type, is an open time-scale strip-chart system on which the rise and fall of the water levels can be reproduced accurately.

There are different types of horizontal-drum recorders, some with spring-wound and others with weight-driven clocks. Most drum recorders have several time and stage-scale ratios available and all record on a strip-chart. Some make only one traverse on a selected time scale, others repeat, and still others record on a continuous roll of graph paper. Those recording on a continuous roll are well-adapted for use in remote locations that are not readily accessible in all weather conditions.

Continuous water-level recorders require protection from the elements, marauders, and the streamflow itself. Protection usually is provided in the form of a *stilling well* for the float and a shelter for the recording

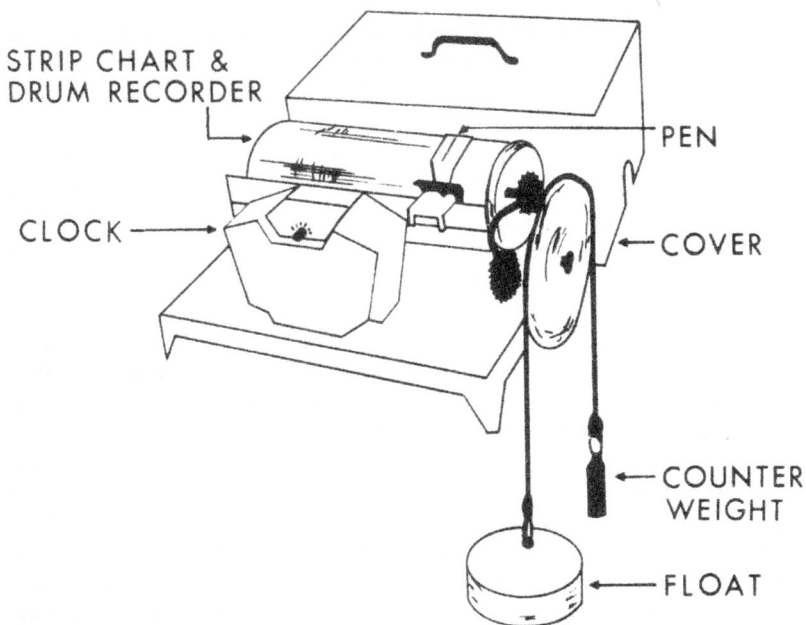

STRIP CHART & DRUM RECORDER

PEN

CLOCK

COVER

COUNTER WEIGHT

FLOAT

Figure 4.3. A schematic drawing of a Stevens Type F continuous water-level recorder.

mechanism. When a stilling well is used, it should be located to one side of the stream channel, so it will not interfere with the normal pattern of the streamflow.

Regardless of how it is measured, the stage of a stream then must be converted to streamflow discharge, either by measuring streamflow velocity at a control section, or by installing a precalibrated structure, for example, a weir or flume, in the stream. Several empirical methods also are available to estimate streamflow discharge values.

4.2.3 Measurement of Streamflow Velocity at a Control Section

The measurement of streamflow velocity at a control section usually involves the use of either a floating object or a standard current meter.

a. Floating object

One of the simplest methods of measuring streamflow discharge is to observe the time required for a floating object that is tossed into the stream to travel a specified distance on the surface of the water. This observation, which is a measure of the streamflow velocity at the surface, then is multiplied by the cross-sectional area of the stream to estimate the streamflow discharge, as follows:

$$Q = VA \qquad (4.1)$$

where Q = streamflow discharge (m^3/s)
V = streamflow velocity (m/s)
A = cross sectional area (m^2)

Although this method is relatively simple, it is not necessarily accurate, because the velocity at the surface of the stream is greater than the average velocity of the stream. The average streamflow velocity generally is assumed to be about 80 to 85 percent of the surface velocity.

b. Current meter

Another method of measuring streamflow discharge is with the use of a *current meter*, an instrument in which a wheel is made to rotate about its axis by the force of the current. The speed of the rotation depends upon the velocity of the water. Importantly, it is necessary to rate a current meter, even when it may appear to be similar to another rated current meter.

To rate a current meter, it is drawn through still water, and the time of travel and number of revolutions of the wheel are observed. The number of revolutions per second and the corresponding velocity in meters per second then are computed. Equations to describe the relationship between the number of revolutions and velocity are developed. The general relationship between the number of revolutions and the velocity of streamflow, which is easily drawn by using a simple linear regression analysis, is represented by the following equation:

$$V = m + k(N) \qquad (4.2)$$

where m = velocity required to overcome the mechanical friction, which approximately is equal to the starting velocity
k = proportionality constant
N = revolution(s)

Two equations frequently are required, one for the higher velocities and one for the lower. The equations then are solved for a range of velocities and a rating table is constructed.

In measuring streamflow discharge with a current meter, the cross sectional area of the stream is divided into a number of vertical sections, and the streamflow velocity is estimated by measuring the average velocity of each section with the current meter (Figure 4.4). The area of each section is determined, and the average streamflow discharge then is computed as the sum of the product of the area and the velocity of each section:

$$Q = \sum_{1}^{n} (A_i V_i) \qquad (4.3)$$

where n = the number of sections

The greater the number of sections, the closer the approximation of streamflow discharge becomes. The number of sections depends largely upon the configuration of the channel and the rate of change in the stage in relation to streamflow discharge. For practical purposes, however, between 10 and 20 sections commonly are used. Regardless of the number of sections, it is important that the:

- Stream depth and streamflow velocity do not vary greatly between measurements.
- The measurements be completed before there is a significant change in the stage. A 10 to 20 cm change in the stage is too much in most cases.

The following guidelines usually are considered in measuring streamflow velocity with a current meter:

- For stream depths greater than 0.5 m, two measurements are taken for each section at 20 and 80 percent of the total depth and then averaged.

- For stream depths less than 0.5 m, one measurement is taken at 60 percent of the total depth.

- For shallow streams less than 0.5 m, a pygmy current meter or similar instrument is used instead of a standard current meter.

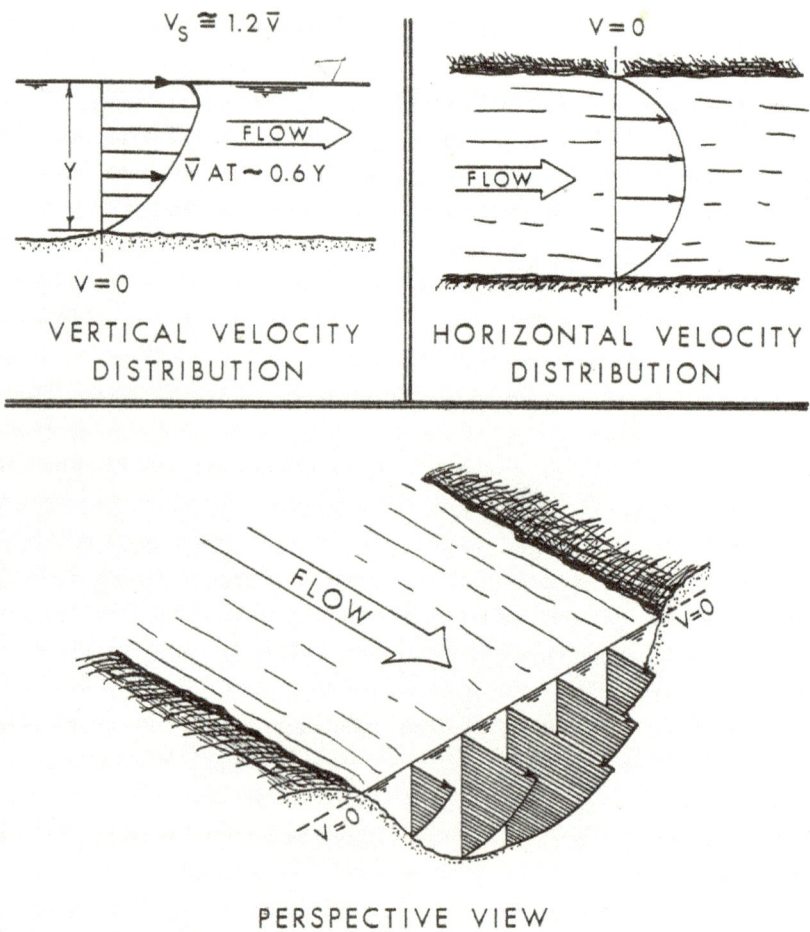

$V_S \cong 1.2\,\overline{V}$

FLOW

$\overline{V}$ AT ~ 0.6 Y

Y

V = 0

VERTICAL VELOCITY
DISTRIBUTION

V = 0

FLOW

HORIZONTAL VELOCITY
DISTRIBUTION

FLOW

V=0

V=0

PERSPECTIVE VIEW

Figure 4.4. Measurements needed to record stream channel cross-section
and streamflow velocity (Brooks et al. 1990).

Measurements of streamflow discharge obtained with a current meter should be tabulated on forms designed for this purpose, as shown in Figure 4.5.

c. Selection of a control section

The most critical aspect of measuring streamflow velocity is the selection of the control section. The control section is the section of the stream for which a *rating curve*, a graph of the stage and streamflow discharge (Figure 4.6), will be developed. To establish a rudimentary rating curve for a staff gage site, only a few measurements of the stage and streamflow discharge may be necessary. However, a complete rating curve for a control section with a continuous water-level recorder may require months or years to establish, as measurements must be taken to represent a range of streamflow discharge conditions.

The stream site selected as a control area should be stable and have a sufficient depth of water for velocity measurements at the lowest streamflow. Other conditions that favor the selection of a stream site are:

- The control site should be in a stream section that is straight for a distance upstream equal to five times the width of the stream, and for a distance downstream equal to two times the width of the stream.
- The stream bed should be smooth and free from vegetative growth, boulders, and other obstructions.
- The stream bed and the banks of the stream should be stable.
- The current should be normal to the control section.
- There should be no large overflow of water at the flood stage.
- The control section should be accessible.

4.2.4 Precalibrated Structures

Whenever possible, a precalibrated structure should be used in measuring streamflow discharge for the total range of flows in a stream. For large watersheds, however, precalibrated structures may not be large enough to measure the high rates of streamflow. In these cases, it may be necessary to use a standard current meter, for example, to measure the high streamflows and a precalibrated structure to measure the lower streamflows.

On small watersheds, less than 1,000 ha in size, precalibrated structures often are used because of their accuracy and convenience. Because of

TABULATION OF STREAMFLOW DISCHARGE DATA
FROM CURRENT METER

LOCATION _____ DATE _____ TIME _____

DISTANCE From Starting Point	WIDTH	DEPTH	METER Location	REVO-LUTIONS	TIME	VELOCITY—		AREA	DISCHARGE	REMARKS
						At Point	Mean In Vertical			
(1)	(2)	(3)	(4)	(5)	(6)	(7)	(8)	(9)	(8.9.10)	
m	m	m	% Depth	number	hr&min	m/sec	m/sec	m²	m³	

Figure 4.5. Example of a tabulation form for measurements of streamflow discharge obtained with a current meter.

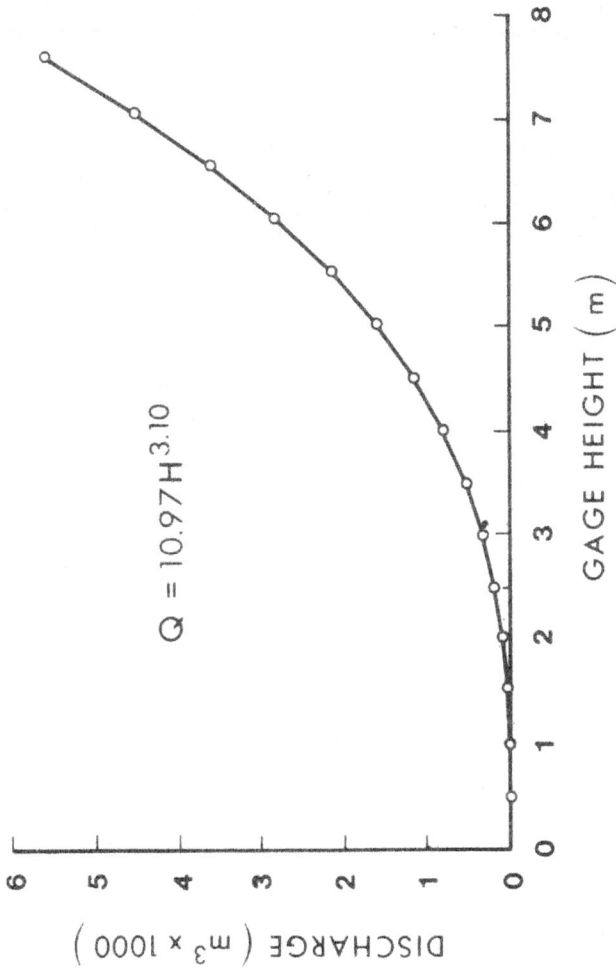

Figure 4.6. Example of a rating curve of stage against streamflow discharge.

$Q = 10.97 H^{3.10}$

DISCHARGE ($m^3 \times 1000$)

GAGE HEIGHT (m)

57

greater accuracy, weirs often are preferred for measuring small watersheds, especially those in which the streamflows become quite low. Where sediment-laden streamflows are common, flumes are more convenient. Weirs and flumes can be constructed from concrete treated wood, concrete blocks, and metals. The notch of a weir generally is a steel blade set into concrete. Flumes often are lined with steel for permanence.

a. Weirs

Weirs are precalibrated structures for which there are mathematical relationships between the *head*, which is the height of water above the *crest*, and the streamflow discharge. The crest is the edge or surface over which the water flows. A record of the head, therefore, can be translated into a record of streamflow discharge.

Weirs include all of the components of a stream-gaging station that incorporates a notch control. The notch can be V-shaped, as shown in Figure 4.7, rectangular, or trapezoidal. An impoundment of water in the stilling basin is formed upstream from the cutoff wall, which acts as a dam, containing the notch. A *stilling well* with a water-level recorder is connected to the weir basin. A shelter is provided to protect the water-level recorder.

The cutoff wall, used to divert the water through the notch, should be tied into bedrock or other impermeable material, so that water does not flow around or under it. In situations where leakage is likely to occur, the stilling basin should be constructed as a water-tight box.

Weirs can be either sharp-crested or broad-crested. A sharp-crested weir has a blade with a sharp upstream edge, so that the passing water touches only a thin edge and clear the rest of the crest. A broad-crested weir has a flat or broad surface over which the water flows. A broad-crested weir generally is used in situations where sensitivity to low streamflows is not critical and where a sharp-crested weir might be damaged by sediment or other debris. Both sharp-crested and broad-crested weirs require that the elevation of the tail water be low enough to eliminate backwater at the crest.

Sharp-crested V-notch weirs often are used where accurate measurements of low streamflows are required. Some of the mathematical relationships between the head and the streamflow discharge for these and the other types of weirs can be found in reference books on hydraulics and

58

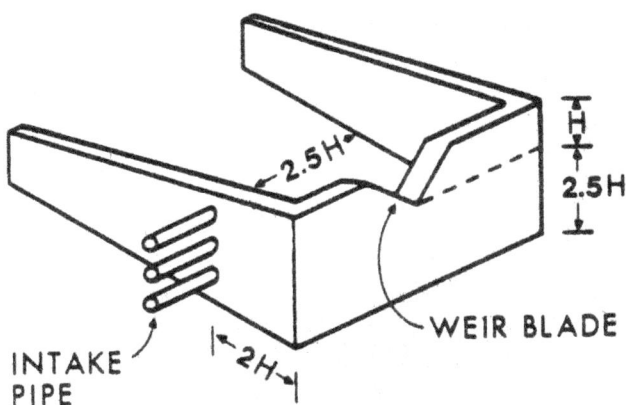

Figure 4.7. Schematic diagram of a weir, showing the minimum requirements for the proper discharge when (H) equals the greatest expected depth for a sharp-crested V-notch weir, with end and crest contractions (Brooks et al. 1990).

hydrology. V-notch weirs can have rectangular sections above to accommodate infrequent high streamflows.

Rectangular weirs have vertical sides and horizontal crests. Their major advantage is their capacities to measure high streamflow rates. Unfortunately, however, rectangular weirs do not allow for precise measurements of low streamflows.

Trapezoidal weirs are similar to rectangular weirs, but they have smaller capacities for the same crest length. The measured streamflow discharges are approximately the sum of the discharges from the rectangular and triangular sections.

b. Flumes

Flumes are artificial open channels built to contain streamflows within designed cross-sectional areas and lengths. In contrast to weirs, there are no impoundments of water, but the height of the water in flumes is measured with stilling wells. The mathematical relationships between the elevation of the water in different types of flumes and the streamflow discharge are found in reference books on hydraulics and hydrology.

HS, H, and HL flumes, developed by the Soil Conservation Service of the U.S. Department of Agriculture, have converging vertical sidewalls that are cut back on a slope at the outlet to provide a trapezoidal projection. (A schematic diagram of an H flume is presented in Figure 4.8). These types of flumes are used to measure intermittent streamflows.

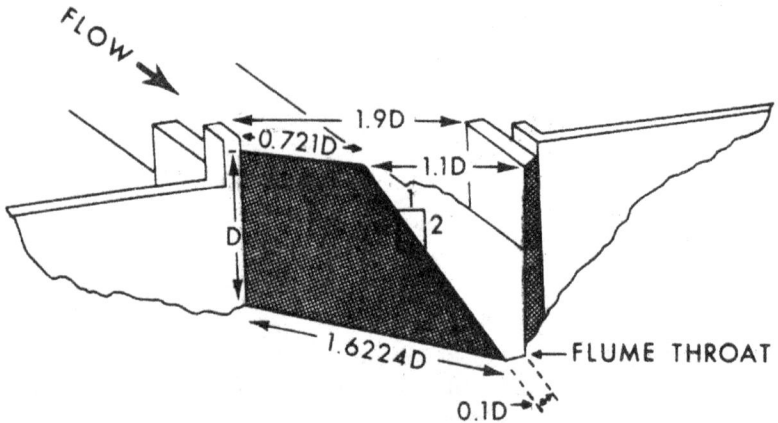

Figure 4.8. Schematic diagram of an H flume, with the dimensions expressed as proportions of the maximum depth (D) (Brooks et al. 1990).

A Venturi flume has a gradually contracting section leading to a constricted throat and an expanding section immediately downstream. The floor is the same grade as the channel. Stilling wells are at the entrance and at the throat, with the difference in the head at the two stilling wells related to the streamflow discharge. This type of flume can be rectangular, trapezoidal, or triangular in shape.

The Parshall flume, which essentially is a modification of a Venturi flume, measures water in an open conduit. It consists of a contracting inlet, a parallel sided throat with a depressed floor, and an expanding outlet, all of which have vertical sidewalls. A Parshall flume is designed to induce a critical flow and hydraulic jump. Two water-level recorders are used when measuring submerged flows, one in the sidewall of the contracting inlet and the other upstream from the lowest point of flow in the throat. When measuring free streamflows, only the upper water-level recorder is used.

The San Dimas flume, developed by the Forest Service of the U.S. Department of Agriculture, measures sediment-laden, rapid streamflows.

It is rectangular in shape and has a sloping floor of 3 percent. The San Dimas flume essentially functions as a broad-crested weir, except that the contraction is from the sides rather than the bottom. Therefore, there is no barrier to cause sediment deposition in the flume. Depth measurements are made in the parallel-walled section at the midpoint. The rapid streamflows keep the flume scoured clean.

c. Selection of a precalibrated structure

The selection of a precalibrated structure to be used depends upon a number of factors, including the:

- Magnitude of the maximum and minimum streamflows.
- Accuracy desired in determining the total streamflow discharges for high flows and low flows.
- Amount and type of sediment and other debris that is expected in the streamflow.
- Channel gradient and cross-sectional area.
- Underlying material.
- Accessibility to the installation site.
- Costs.

4.2.5 Empirical Methods

Several empirical methods are available to estimate streamflow discharge values. Two commonly employed empirical methods for estimating streamflow velocities at known stages are the Manning and the Chezy equations. The estimates of streamflow velocities obtained from solutions of these equations can be converted to streamflow discharge values by multiplying the estimates by the cross-sectional area of the stream.

a. Manning equation

The Manning equation is:

$$V = \frac{1}{n} [R^{2/3}S^{1/2}] \tag{4.4}$$

where n = Manning resistance coefficient

R = hydraulic radius (m) = $\frac{A}{P}$

P = wetted perimeter

S = slope of the water surface (m/m)

b. Chezy equation

The Chezy equation is:

$$V = C [RS]^{1/2} \qquad (4.5)$$

where C = Chezy resistance coefficient

Equations (4.3) and (4.4) are similar in structure. The relationship between the two resistance coefficients is:

$$C = \frac{1.81}{n} [R^{1/6}] \qquad (4.6)$$

Equations (4.3) and (4.4) also are applied in a similar fashion. The hydraulic radius and the slope of the water surface are obtained from cross sectional and stream bed slope data collected in the field (Figure 4.9). The Manning resistance coefficient, which can be converted to the Chezy resistance coefficient when the Chezy equation is used, can be estimated from values in Table 4.1.

4.3 DATA COLLECTION AND PROCESSING

4.3.1 Time and Frequency of Data Collection

A staff gage site generally is visited at regular intervals specified by the watershed manager. A crest gage records only the highest stage reached between visits, so the visits may have to be more frequent when a crest gage is installed. During periods of flooding, a staff gage or crest gage site may be visited once or twice daily.

For the control sections with continous water-level recorders, the clocks should be geared to make one revolution every 6, 12, or 24 hours, depending largely upon the rapidity with which the stage changes. All clocks generally run 8 days with one winding, regardless of the gears used.

The control sections with continuous water-level recorders should be visited at least once a week during periods of little or no rainfall, and after each period of stormflow. These visits, which can coincide with those to rain gage locations, should be scheduled in the morning of the day after the end of a rainfall event.

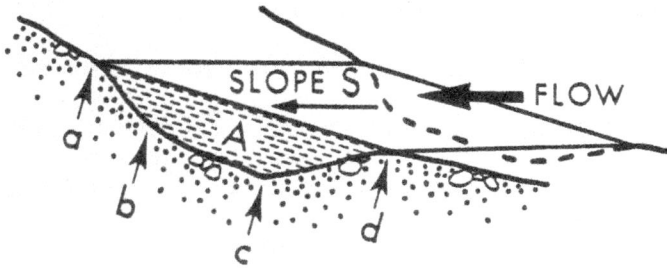

Figure 4.9. Stream channel section showing the slope or gradient of the streambed, the wetted perimeter (P), which is line A-B-C-D, and cross-sectional area (A) (Brooks et al. 1990).

Table 4.1. Values For The Manning Resistance Coefficient (Gray 1970).

Type of Channel	Minimum	Average	Maximum
Natural streams			
Mountain streams, no vegetation in the channel, bed of gravels and cobbles	0.030	0.040	0.050
Sluggish stream reaches with vegetation and deep poles	0.050	0.070	0.080
Flood plains			
Pastureland, no brush, short grasses	0.025	0.030	0.035
Scattered brush, dense vegetation	0.035	0.050	0.070
Dense woody vegetation, straight channel	0.110	0.150	0.200

63

4.3.2   Sources of Errors in Data Collection

Sampling errors generally are not considered in measurements of streamflow discharge. However, errors in a rating curve, including the errors in taking the measurements that are required for the development of the rating curve and the statistical variability in the curve itself, should be considered. Instrument-related errors can occur, especially when a continuous water-level recorder is employed to determine the stage. Proper maintenance, therefore, is important.

4.3.3   Data Processing

The processing of data collected at a staff gage site is relatively simple. Observations made of the stage are recorded on appropriate forms, and then brought into the office where the corresponding streamflow discharge values are determined from a rating curve. The processing of data obtained from a continuous water-level recorder is more detailed, however.

Processing of data from continuous water-level recorders should begin immediately after the charts and the supporting field notes are brought into the office. The processing usually is done in three separate operations, possibly by three different individuals. Initially, the field notes and other information that may be required in the compilation of the data are transferred onto the charts. Secondly, the data from the charts are tabulated in sufficient detail to ensure that the subsequent computations will meet the desired level of refinement. The final step is the computation of streamflow discharge and total amounts of streamflow.

a.   Chart annotations

For charts that cover periods during which no streamflow occurred, it is necessary to show only the chart number and dates. No other notes are required, as the main purpose of these charts is to show the continuity of the streamflow records.

For charts that cover periods of streamflow, the following procedure should be followed:

- Enter the chart number and dates.
- Show the time zone.
- Enter the dates, watch time, index pointer, chart line, and stage reading at times of placement, inspection, and removal.

- Indicate the time and stage-scale used.
- Check the placement and removal marks with the watch time. If they do not agree within 2 minutes, a time correction should be made.
- Enter the correct times of rise, noons, and midnights subsequent to the rise and other points necessary to simplify the compilation of the records.
- Check for discrepancies between the chart line and the index pointer, and also for failure of the pen to reverse at the edges of the printed part of the chart. If the pen reverses below the limits of the printed part of the chart at about the same magnitude at both the upper and lower reversals, apply a constant correction to each traverse. This correction is plus for the traverse upward across the chart, while it is minus for the downward traverse. In cases where the lower reversal is correct and the upper reversal is short, a graduated correction theoretically would be correct. Note any correction that must be applied to the stage.
- Show the corrected stage values at all peaks, troughs, and reversals of the actual record.
- Enter notes to show when the stage is rising or falling.

b. Tabulation of data

The tabulation of the data from a continuous water-level recorder chart should supply sufficient information to provide for accuracy in the determination of the streamflow discharge and total amounts of streamflow. The method to be used in the calculation of streamflow discharge will influence the frequency with which the data points must be selected. Appropriate guidelines for selecting data points should be established before tabulation begins. Experience should indicate the refinement that may be necessary to fit a particular situation.

The record from a continuous water-level recorder chart, corrected in accordance with the instructions presented above, should be tabulated on appropriately designed forms, as illustrated in Figure 4.10. Entries on these forms should be made as follows:

- Fill in all data indicated. The file number should include the site designation, the year, and the sheet number. Once again, the dates shown should include those periods with no streamflow to give a continuity to the record.
- Show the period since the last streamflow, if it is more than one calendar day, on the first line of the form.

TABULATION OF RUNOFF DATA

LOCATION: _____   GAGE DATUM ELEVATION OF POINT OF ZERO FLOW: _____ m
DRAINAGE AREA: ____ ha ____ km²   GAGE DATUM ELEVATION OF POINT OF ZERO PONDAGE: ____ m
DATE: _____ TO _____   GAGE DATUM ELEVATION OF FLOAT AT REST: _____ m
RATING TABLE: _____   GAGE DATUM ELEVATION OF LOWEST INTAKE: _____ m

DATE & TIME (1)	TIME INTERVAL (2)	GAGE HEIGHT (3)	RATE OF CHANGE IN STAGE (4)	PONDAGE CORREC- TION (5)	OBSERVED DISCHARGE (6)	RATE OF RUNOFF (Discharge corrected for pondage) (7)	(8)	AVERAGE RATE OF RUNOFF FOR TIME INTERVAL (9)	TOTAL RUNOFF For time interval (10)	Accumu- lated (11)	REMARKS (12)
hr & min	min	m	m/min	sec-min	sec-min	sec-min	cm/hr		cm	cm	

- Enter the month, day, and time in column (1) to correspond to the stage values selected in column (3).
- Enter the stage values obtained from the continuous water-level recorder charts in column (3). In general, tabulate stages at any noticeable break in the pen trace, and at midnight in the case of streamflows involving more than one day.

c. Calculations

A constriction in a stream channel often will cause a certain amount of ponding upstream. The effect of this ponding upon the streamflow discharge should be considered in selecting the procedure for calculating the streamflow discharge and total amount of streamflow. Careful selection of the control section will minimize the effect of ponding.

The general procedures that are outlined below are intended for use in situations where the ponding above the control section, if present, in which case it is recorded in column (5), is so small that the streamflow peaks are not affected appreciably:

- On the tabulation form, enter in column (2) the difference (in minutes) between consecutive entries in column (1). The summation of the numbers in column (2) should equal the total time-lapse in column (1).
- Enter in column (8) the streamflow discharge taken from the rating table for each stage value in column (3).
- Enter the average streamflow discharge for each time-interval in column (9). It is the average between successive streamflow discharge values from column (8).
- For the total amount of streamflow for the time-interval in column (10), multiply the values tabulated in column (9) and the corresponding time-interval (in minutes) in column (2), and divide by 60.
- The accumulated summation of the values in column (10) are entered in column (11). (Carry the entries in columns (9), (10), and (11) to four decimals.)
- In column (12), enter notes to show the beginning of rise, maximum stage and maximum streamflow discharge, end of streamflow, daily totals, and any other pertinent notes.

4.3.4 Innovations in Data Processing

In recent years, a number of innovations in data processing have evolved to lessen the tasks of data collection and processing. Two of these innovations, digital recorders and telemetering systems, are described briefly in the following paragraphs.

a. Digital recorders

Many watershed managers consider a punched tape, rather than a graphic format on a chart, to be the most desirable format for data collection and processing. Digital recorders, such as the Stevens Series 7000 Digital Recorder, for example, convert angular shaft positions in a continuous water-level recorder into coded digital information, and record this information periodically as a pattern of punched holes. Power usually is supplied by batteries, although alternating electrical current also can be used if it is available.

Four digits of information generally are available. For water-level measurements, this can be 0 to 99.99 m (where the smallest increment equals 1 cm) or 0 to 9.999 m (where the smallest increment equals 1 mm). The coding is accomplished by four encoding drums, one for each digit, that have a pattern of "ridges" or "valleys." The "ridges" on each drum are assigned weighted values of either 8, 4, 2, or 1. These values are added together to obtain a decimal number from each of the four drums.

The punching of the tape is controlled by a timer. Pre-selected intervals of 5, 15, 30, or 60 minutes are available, using a different cam for each of the different intervals of time. The cam on the timer essentially corresponds to a minute-hand of a watch, in that it makes one revolution each hour. If the cam has one drop-off point, the digital recorder will punch hourly, if it has two drop-off points, it will punch every 30 minutes, etc. During the punching operation, the tape is pressed against the punch pins. These pins are backed up by the encoding drums, so that those pins which rest against a "ridge" will punch a hole in the tape.

Digital recorders generally are available with a variety of options and accessories, and can be used for many different applications. For water-level measurements, digital recorders can be used with either a float or bubble gage. Digital recorders for rainfall measurements also are available, as are digital recorders to punch in water quality parameters.

Electronic translators are used to convert the punched tapes into suitable input data for digital computers, in which the desired calculations and data summaries can be made.

The instrument manual provided by the manufacturer of the digital recorder selected for use should be referenced for detailed information on the operating procedures, maintenance, and calibration.

b. **Telemetering systems**

Telemetering systems can be used for data handling of specific variables, including streamflow discharge. The selection of a telemetering system depends largely on the other equipment already available, the location, and the financial resources available. In general, two basic types of telemetering systems are used in watershed management, specifically, analog systems and digital systems.

Analog systems provide an output that is proportional to the input. To illustrate, on a float operated water-level transmitter, the float rises and falls with changes in the stage. This action, in turn, is transmitted through a set of gearing and synchronized motors to a receiver, where it is displayed and recorded. If the receiver is a graphical recorder, the graph normally is recorded at a reduced scale. The recording, therefore, becomes a visual "analogy" of the water-level change.

Digital systems transmit information as a digitally-coded message. Using the float mentioned in the analog description, the float rises and rotates the float pulley on the transmitter. Internally, this input is converted into pulses or contact closures. Receivers are available to decode the "digital" transmissions and to register the information.

The type of telemetering system selected for use will be dependent, in large part, on the communication channel available. Although the most accurate and often least expensive systems are position motor equipment, these are limited to a maximum distance of a few km and must use an alternating electrical current circuit. For longer distances, impulse systems may be more appropriate, using almost any kind of two wire physical circuit, such as telephone lines or radio equipment. Normal transmission with these latter systems is about 80 km. However, with repeater stations, this distance can be extended indefinitely.

4.3.5 Missing Data

On occasion, a record of the total amount of streamflow from a specific rainfall event is lost because of the malfunctioning of a continuous water-level recorder. In such instances, an approximation of the missing record may become necessary.

It is possible to estimate the streamflow discharge through solutions of either the Manning equation or the Chezy equations, assuming that the required inputs are available. Estimates of streamflow velocities initially are obtained through solutions to these equations, after which the streamflow discharge can be estimated.

a. Rational runoff method

Another approach to estimating peak streamflow discharge involves the application of the so-called *rational runoff method*, which requires data on rainfall intensities and watershed characteristics. In this method, it is assumed that a storm of a uniform rainfall intensity covers a watershed. The resulting streamflow will increase as flow from more remote parts of the watershed reaches the outlet. When the whole watershed is contributing, a steady state is attained and streamflow discharge becomes a constant maximum. The time required for this steady state to be reached is called the *time of concentration* for the watershed. After this time, the streamflow discharge is a fixed proportion of the rainfall intensity, and it is equal to:

$$Q_p = 0.278 \, [CIA] \tag{4.7}$$

Q_p = peak streamflow discharge (m^3/s)
C = rational runoff coefficient
I = rainfall intensity (mm/hr)
A = size of watershed (km^2)

Values for the rational runoff coefficient are listed in Table 4.2. These values, which generally reflect soil type, topography, surface roughness, vegetation, and land use, are assumed to remain constant during and between large rainfall events on a specific watershed. The rational runoff method generally should be used only for watersheds of less than 1,000 ha.

Table 4.2. Values of the rational runoff coefficent (American Society of Civil Engineers 1969).

Watershed Characteristics	C
Forests	0.10
Pastures	0.15
Cultivated lands	0.20
Loamy soils without impeding horizons	
Forests	0.30
Pastures	0.35
Cultivated lands	0.40
Heavy clay soils or those with shallow impeding horizon; and shallow soils on bedrock	
Forests	0.40
Pastures	0.45
Cultivated lands	0.50

The time of concentration, which is the time required for water to travel from the most distant point on a watershed to the watershed outlet, can be estimated by:

$$T_c = 0.02 \left[\frac{L^{1.2}}{H^{0.4}} \right] \qquad (4.8)$$

where T_c = time of concentration (hr)

L = length of the watershed along the main stream channel, from the most distant point to the outlet (m)

H = difference in elevation on the watershed between the most distant point to the outlet (m)

The assumption that the rainfall intensity is uniform on the entire watershed for the period equal to the time of concentration is seldom met, especially under natural conditions. To apply this method, therefore, rainfall intensity-duration values that are associated with an acceptable risk are employed.

b. Synthesis of a hydrograph

The synthesis of a hydrograph to represent a missing stormflow event often is required. Unfortunately, this is not an easy task without some knowledge of the stage at peak flow, an estimate of initiation and cessation times of the stormflow event, and the time to peak.

The stage at peak flow generally is indicated by the "high-water mark" for the stormflow. The time of initiation and cessation of the stormflow event, and the time to peak can be approximated only, although experience on the hydrological behavior of the watershed in question to specific rainfall events can be helpful in making these estimates. With this basic information, a simple hydrograph in the shape of a triangle or a composite of triangles can be synthesized to represent the missing stormflow event. Subsequently, this simple hydrograph can be refined to a more complex geometric shape or statistical distribution to simulate a specific set of rising limb and recession limb characteristics.

The most common method of prediction of the hydrograph involves the construction of a *unit hydrograph*, as described in Section 4.5 of this manual.

4.4 DATA ANALYSIS

4.4.1 Double Mass Analysis

If it is suspected that the relationship between the streamflows measured at two or more control sections has changed, for whatever reason, a double mass analysis of the data can be undertaken to check for the homogeneity of the records. The general procedure followed in making this analysis is similar to that outlined in checking the accuracy and consistency of rainfall measurements taken at one or more locations (see Section 2.5.2 of this manual).

4.4.2 Mass Curve Analysis

In some situations, it is required to use water downstream at uniform rates that are greater than the minimum streamflow discharge. It often is necessary, therefore, to provide storage space in which water can be impounded during periods of high streamflow for use during periods of low streamflow. The determination of the required storage capacity for the uniform use rates is a problem that can be solved through an analysis of mass curve of the streamflow discharges during a selected period of streamflow record.

A mass curve is a graphical representation of the accumulated streamflow discharge in relation to time. The difference in the ordinates at the ends of a small segment of a mass curve is the volume of water discharged in the time period shown by the corresponding difference in the abscissas. Therefore, the slope of a mass curve at any point is equal to the rate of streamflow discharge at that time. In other words, the slopes of a mass curve are equal numerically to the corresponding ordinates of the hydrograph.

Suppose, for example, that a mass curve for a specific stream has been plotted. Then, if a mass curve for a uniform use rate is plotted, it is a straight line, such as the line are in Figure 4.11, which is drawn so that its slope is 2 m^3/s. Two other lines, parallel to are and tangent to the mass curve at points b and c, also are shown. During the time periods (a to b) and (c to b'), when the slope of the mass curve is greater than that of the straight line are, the streamflow discharge is greater than 2 m^3/s, while the streamflow discharge is less than 2 m^3/s from b to c. If the storage facility is assumed to be filled to capacity at a, then the line abdb' becomes a mass curve of outflow, with a minimum outflow of 2 m^3/s. As any point on this line represents the total outflow to that time, and because the point on the mass curve of inflow abcdb' found on the same vertical line represents the total inflow up to that same time, the vertical distance between these two mass curves represents the amount that the storage facility is drawn down at that time. Importantly, the greatest of these drafts is the size of the reservoir needed during this period to provide a minimum flow of 2 m^3/s.

It follows then that the greatest vertical distance (s) between bd and bcd, which occurs at c, is the storage required to maintain a uniform use rate of 2 m^3/s during the low streamflow period from b to c. The largest value, such as s for the entire period of record, is the minimum size of the storage facility which would provide this uniform use rate.

4.4.3 Return Interval

The frequency of streamflow events can be described in terms of either probabilities of occurrence or *return interval*, the latter also is called the *recurrence interval*. *Exceedance frequency*, the frequency with which a

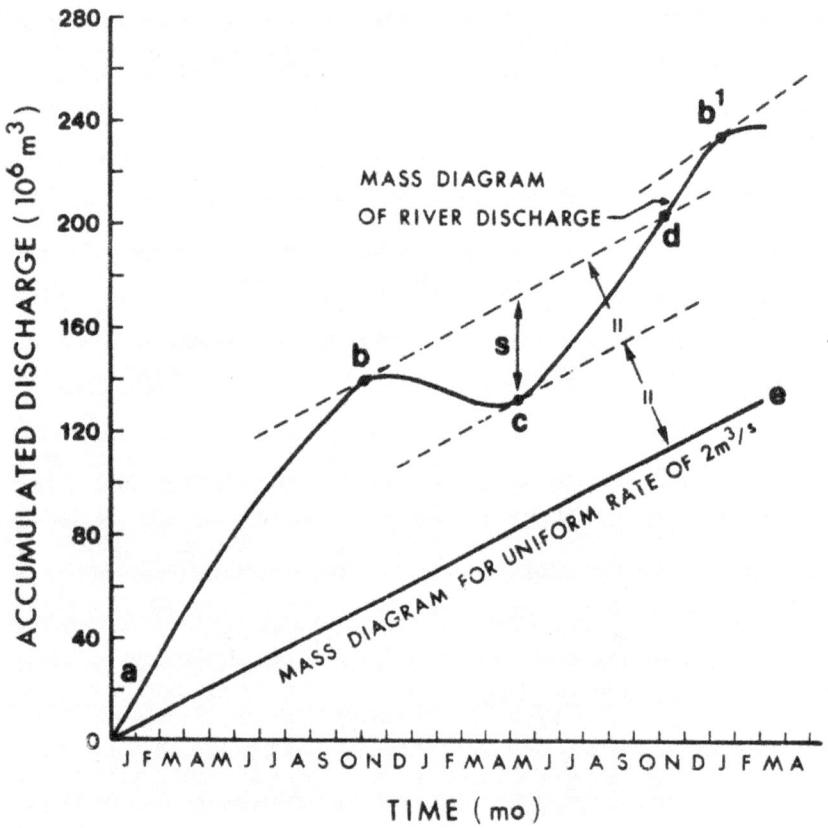

Figure 4.11. Example of a mass curve for a stream (Wisler and Brater 1965).

streamflow event of a specified magnitude is equalled or exceeded, can be determined from a frequency analysis, specifically with a frequency curve. A frequency curve is simply an expression of streamflow data on a probabilistic basis. In developing a frequency curve, the following points must be considered:

- The streamflow characteristics, for example, instantaneous peak flo ws or mean daily peak flow, need to be identified.

- The data utilized in the construction of the frequency curve must represent a measure of the same aspect of the streamflow event. In other words, instantaneous peak flows cannot be analyzed together with mean daily peak flows.
- The streamflow data being analyzed should be controlled by a uniform set of hydrological factors. That is, natural streamflow records can not be analyzed along with flows that have been modified by a storage facility.
- When only a few years of streamflow records are available, a *partial duration series* analysis can be used, rather than an *annual series* to define the frequency curve. With a partial duration series, all of the independent events above or below a specified base level are used in the analysis, while only the extreme event is selected from the streamflow record for each year with an annual series.

  The frequency curves based on the two approaches are interpreted differently, however. For a partial duration series, the exceedence frequency for a specified magnitude is the number of "events" that will be exceeded in a 100 year period. For an annual series, the interpretation is the number of "years" that will be exceeded in a 100 year period.

Once the appropriate streamflow data sets have been defined, a frequency curve can be developed by either graphical or analytical methods.

a. Graphical method

The graphical method involves the calculation of plotting positions, or probabilities, for ranked events, and then drawing the frequency curve through the data points. A commonly used formula to calculate the plotting positions for a peak flow frequency curve is:

$$P = \frac{m - 0.3}{N + 0.4} \qquad (4.9)$$

where P = the plotting position, or probability, for the event ranked m in N number of years of streamflow record

The plotting position for the "least severe" events (probability < 50 percent) can be calculated by:

$$P = \frac{2m - 1}{2N} \tag{4.10}$$

The graphical method does not assume a specific statistical distribution. Because the frequency curve is drawn by hand, it is possible to weigh one portion of the curve more than another. It is recommended that the plotting positions determined through the graphical method be plotted even if the analytical method is used to define the frequency curve.

b. Analytical method

The analytical method requires that the data points follow a theoretical frequency distribution. For peak flow frequency, the log Pearson type III distribution often is recommended. This method requires the steps:

- The data are transformed by taking the base logarithms of the peak flow discharges.
- The mean peak discharge, the first moment, is calculated; this corresponds to the 50 percent probability of exceedence.
- The standard deviation, the second moment, is calculated; this represents the slope of the frequency curve plotted on log-probability graph paper.
- The skew coefficient, the third moment, an index of non-normality, is calculated; this represents the curvature of the frequency curve.
- Adjustments then are made in cases of small numbers of events. The skew coefficient usually is unreliable for limited records. A regional skew coefficient is recommended for less than 25 years of streamflow record.
- The frequency curve for the observed annual peak flows then is determined for selected exceedence probabilities by:

$$\log Q = x + [K_s S] \tag{4.11}$$

where Q = observed annual peak flows
x = mean peak streamflow discharge

K_s = factor that is a function of the skew coefficient and a selected exceedence probability

S = standard deviation

- Confidence limits are calculated and plotted for the frequency curve.

The analytical method has a number of advantages over the graphical method, including:

- The same curve always is calculated from the same data sets, which makes it more objective and consistent in analysis.
- The reliability of the frequency curve can be calculated with confidence intervals.
- The calculation of statistics by a regional analysis, which is a statistical approach in which a generalized equation, graphical relationship, or map is developed for estimating streamflow measurements on an ungaged watershed, allows the development of a frequency curve on the ungaged watershed.

4.4.4   Highflow and Lowflow Regimes

Many of the land use practices on watersheds result in a change in the streamflow regimes, for example, an extension of the periods of high streamflows or an increase in the magnitude of the streamflow in periods of low flow. In analyzing these and other changes in the streamflow regimes, stormflow events representing either highflows or lowflows are chosen for a frequency analysis (see Section 4.4.3 of this manual).

The criteria employed in the selection of these stormflow events, although generally arbitrary, should be based on a specific purpose, such as studying the effects of timber harvesting activities on the occurrence of either highflows or lowflows of a specified magnitude. A frequency analysis of highflow and lowflow regimes can be helpful in comparing the hydrological performance among a set of watersheds. In doing this, questions such as whether or not the effects of timber harvesting activities on highflows or lowflows are general in nature can be answered.

4.5   UNIT HYDROGRAPH

One of the more common methods of hydrograph prediction involves the construction of the *unit hydrograph*, which can be defined as the hydrograph of 1 unit of stormflow that results from a rainfall event that occurs at a fairly uniform intensity within a specific period of time. In essence, the unit hydrograph method is simply a "black box" representation that empirically relates stormflow from a watershed to a specified duration of rainfall. No attempt is made to simulate the hydrological processes involved in the flow of water through the watershed.

4.5.1 Theoretical Basis

The principles that constitute the theory of the unit hydrograph include:

- A unit hydrograph is a representation of the surface runoff component of streamflow that results from a relatively short, intense rainfall event, called a *unit storm*.
- A unit storm is a rainfall event of such duration that the period of surface runoff is not significantly less for a rainfall event of short duration. Its duration is equal to or less than the "period of rise" of a unit hydrograph, which is the time from the beginning of surface runoff to the peak. The period of surface runoff is approximately the same for all unit storms, regardless of their intensity.
- A "distribution graph" is a graph that has the same time-scale as a unit hydrograph and ordinates which are the percent of the total surface runoff that occurred in successive, arbitrarily chosen, uniform intervals of time. The most important concept involved in the unit hydrograph theory is that all unit storms, regardless of their magnitudes, result in nearly identical distribution graphs. Therefore, once a distribution graph has been derived from a watershed, it serves as a means of converting any expected volumes of surface runoff into a hydrograph of streamflow discharge.

These relationships are not necessarily true, strictly speaking, although the error is of little consequence from a practical viewpoint.

4.5.2 Development of a Unit Hydrograph

The unit hydrograph concept can be best understood through an example, in this case, the *isolated storm method* of developing a unit hydrograph, as outlined in Figure 4.12 and described below.

- Streamflow records for the watershed in question are studied initially, selecting a single-peaked, isolated hydrograph which resulted from a rainfall event of short duration and uniform intensity.- Once a hydrograph has been selected, the stormflow volume is determined by separating the more uniform baseflow

A DETERMINE VOLUME OF STORMFLOW FROM AN ISOLATED STORM:

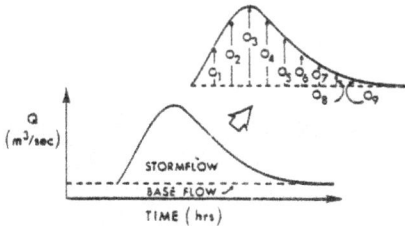

$$\left[\frac{STORMFLOW \ (m^3/sec \times sec_d)}{AREA \ (m^2)}\right] \left[\frac{1000 \ mm}{m}\right] = \begin{array}{c} mm \ of \ STORMFLOW \\ (Q_S) \end{array}$$

Where sec_d = Duration of Stormflow

B DETERMINE ORDINATES FOR THE UNIT HYDROGRAPH (1mm Stormflow);
DIVIDE EACH ORDINATE ABOVE (Q_1, Q_2 ...)
BY THE mm OF STORMFLOW.
THE NEW ORDINATES ARE FOR 1mm
OF STORMFLOW VOLUME:

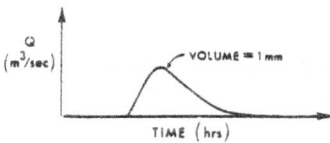

C DETERMINE RAINFALL FOR THE WATERSHED:

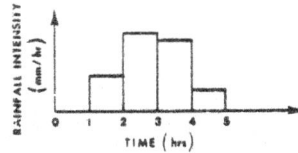

D ESTIMATE DURATION OF EFFECTIVE RAINFALL

1) LOSSES = TOTAL RAINFALL
   - TOTAL STORMFLOW DEPTH

2) AVERAGE HOURLY LOSS = $\dfrac{RAINFALL - STORMFLOW}{RAINFALL \ DURATION}$

3) DURATION EFFECTIVE RAINFALL = 2 HOURS

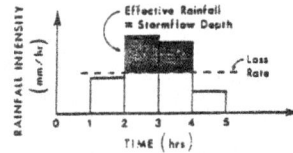

Figure 4.12. Development of a unit hydrolograph by the isolated storm method (Brooks et al. 1990).

component from the rapidly changing stormflow component, as described in Section 4.6.1 of this manual.

- The stormflow then is converted into mm of depth for the watershed area.

- The average rainfall depth on the watershed that caused the stormflow hydrograph is determined.
- The time distribution of the rainfall then is determined and the total rainfall amount is compared to the stormflow volume.
- The difference between the average rainfall depth and the stormflow volume, which sometimes is called the *effective rainfall*, is considered to represent "losses." The magnitude of the losses is a function of the antecedent moisture conditions on the watershed. A uniform rate of loss has been assumed in Figure 4.12.
- The duration of the effective rainfall then is determined, which in Figure 4.12 is 2 hours.
- The ordinates of the stormflow hydrograph are divided by the mm of stormflow to obtain the ordinates that correspond to 1 mm of stormflow.

The duration of the effective rainfall defines the unit hydrograph, for example, a unit hydrograph developed from an effective rainfall of 6 hours in duration is a 6-hour unit hydrograph. The unit hydrograph shown in Figure 4.12 is a 2-hour unit hydrograph.

A unit hydrograph also can be developed from a multiple-peaked stormflow hydrograph, although these more complex stormflow events should be analyzed with computer assistance.

4.6 STORMFLOW RESPONSE

The unit hydrograph is a widely used method of estimating the stormflow response of a watershed. Simpler approaches also can be used, however, such as the *streamflow response*, as described below.

The stormflow response of a watershed can be characterized by calculating the average ratio of the stormflow volume, also referred to as the *quickflow volume*, to the rainfall volume for different periods of stormflow. Once again, the stormflow volume is that portion of the hydrograph that responds quickly to a rainfall event and is determined by separating the baseflow, also called *delayed flow*, from the total streamflow during the storm event.

80

Through comparisons of stormflow response factors for watersheds of different vegetative covers, soils, and land uses, the respective flooding potentials of the watersheds can be estimated. In addition, a stormflow response factor can be used to develop an estimate of peak stormflow discharge, when a consistent relationship between the stormflow volume and the peak is present.

4.6.1 Baseflow Separation

The baseflow component of a hydrograph must be separated from the total streamflow in calculating a stormflow response factor. Unfortunately, there is no one method of baseflow separation, because the separate pathways of flow of excess water on and through a watershed cannot be related directly to a hydrograph. A hydrograph, instead, represents the integrated responses of all pathways of flow. This fact is not necessarily a problem, however, as the baseflow contribution to a hydrograph is typically a small portion of the stormflow.

- Graphically separate the baseflow from the stormflow for several stormflow hydrographs, as illustrated by methods I and II in Figure 4.13.
- After examining several stormflow hydrographs, determine if there is a consistent relationship that can be expressed through the use of either method I or method II.
- In method I, the separation line I is drawn between the point of hydrograph rise to the point of cessation of the stormflow. Determine if the separation line I yields a consistent rate in terms of $m^3/s/km^2$.
- In method II, the separation line II is drawn initially between the point of hydrograph rise to a point on the recession limb of the hydrograph that is defined by N days after the peak, where:

$$N = A^{0.8} \tag{4.12}$$

where A = watershed area (km^2)

4.6.2 A Stormflow Response Relationship

Streamflow analyses of watersheds in the southeastern portion of the United States have resulted in the following equation of stormflow response, expressed in English units:

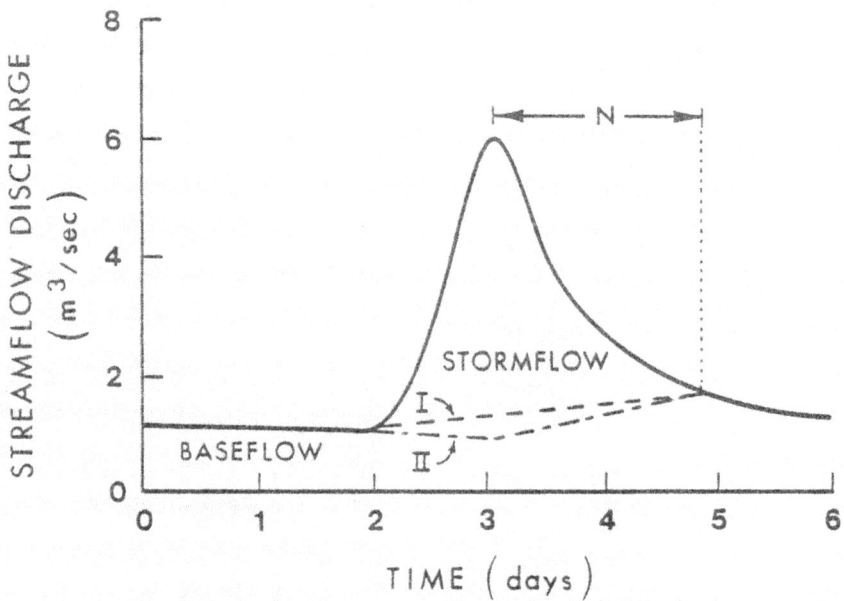

Figure 4.13. Methods of separating the baseflow from the stormflow (Brooks et a. 1990).

From the point on the recession limb of the hydrograph, the separation line continues to the points of cessation of stormflow.

$$Q_s = 0.022[RSP^2] \qquad (4.13)$$

where Q_s = stormflow volume (mean depth in inches)

$$R = \sum_1^n [Q_sP]/n, \text{ for n observations and } P \geq 1 \text{ in.}$$

S = a "sine-day" factor

$$= SIN\ [360]\ [\ \frac{Day\ No}{365}\ ] + 2, \text{ where Day 0 = November 21}$$

P = rainfall (mean depth in inches)

The "sine-day" factor in equation (4.13) is an approximation of the antecedent moisture conditions on the watershed, presented as a seasonal coefficient.

The relationship between stormflow volumes and peak stormflows above the baseflow component of the hydrograph was:

$$Q_p = 25[RS^{0.5}P^{2.5}] \qquad (4.14)$$

where Q_p = peak stormflow discharge ($ft^3/s/mi^2$)

While the example presented above is representative of specific conditions in the United States, the mathematical forms of the equations may be helpful in the development of similar relationships in the ASEAN region.

CHAPTER FIVE

Sediment

5.1 SOIL PARTICLE DETACHMENT AND TRANSPORT

Dislodgement of soil particles through the energy of rainfall impacting on the soil surface is a primary agent of erosion on watersheds. However, only a portion of the dislodged soil particles is passed through and off a watershed as sediment in streamflow. Sediment, therefore, is the product of erosion, whether the erosion occurred as surface, gully, or mass soil erosion. Sediment is deposited at the base of hillslopes, in flood plains, and within stream and river channels. The rate at which sediment is discharged into the ocean is generally less than the rate of erosion on upland watersheds and is dependent largely upon the physical characteristics of the stream channels.

The term *sediment* generally refers to the soil particles that are transported by the streamflow. *Sedimentation*, in turn, is the process of deposition of the transported soil particles from the water.

5.1.1 Sediment Yield

The *sediment yield* of a watershed is the total sediment outflow from the watershed, as measured at a defined point in the stream channel for a specified period of time. Sediment yield is determined by relating the measured sediment to the streamflow discharge, or through a survey of deposited sediments in a reservoir. Streams that discharge large quantities of sediment on an annual basis are those flowing from watersheds undergoing active geologic erosion or being subjected to improper land use practices.

5.1.2 Process of Sediment Movement

The sediment discharge of a stream is the mass rate of transport through a specific cross-sectional area of the stream. It generally is measured in terms of mass per second per meter of stream width. A portion of the sediment discharge consists of fine soil particles, such as silt and clay, which are transported in suspension. This portion is referred to as the *suspended load* or *wash load*. Another portion of the sediment discharge, called the *bed load*, consists of sands, gravels, or cobbles. The bed load is transported along the bottom of the stream channel by traction, sliding, or saltation, as shown in Figure 5.1.

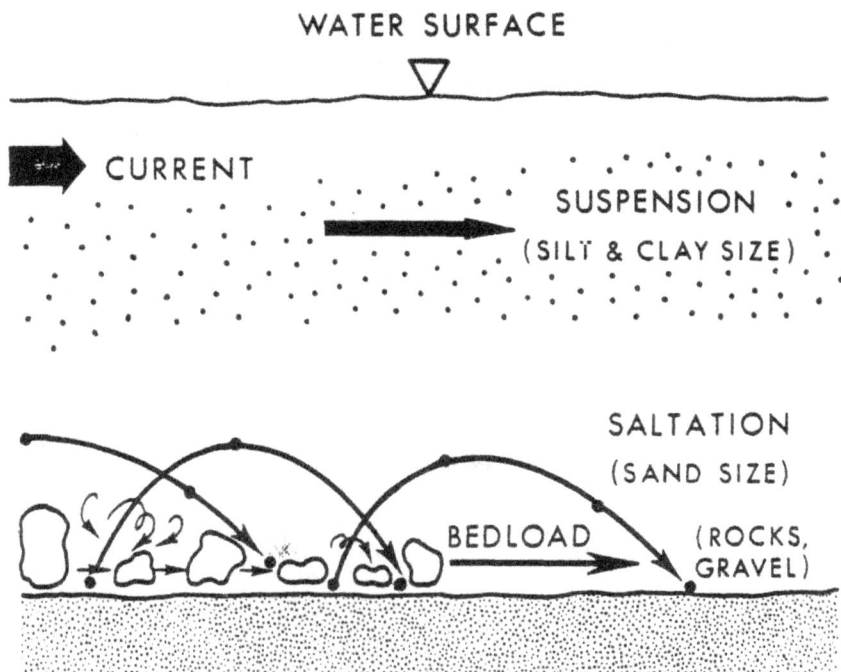

Figure 5.1. Transportation of soil particles in flowing water (Brooks et al. 1990).

The sediment discharge of a stream is dependent, in large part, upon the interrelationships among the supply of soil particles to the stream channel, the physical characteristics of the sediment, the characteristics of the stream channel, and the rate of streamflow discharge. The supply of soil particles and the rate of sediment discharge depends largely upon the climate, geology, soils, topography, vegetation, and land use practices. Stream channel characteristics of importance include the morphological stage of the stream channel, the roughness of the bottom of the stream channel, and the steepness of the stream channel. The state of weathering of the geological materials and soils on the watershed determine the physical characteristics of the sediment.

Interrelationships among the above factors determine the amount and type of sediment supply and the energy that is available to transport the soil particles. In situations where the energy is greater than the sediment supply, *stream channel degradation* occurs. On the other hand, when the sediment supply is greater than the energy, *aggradation* takes place within the stream channel. The relationship between the transport capability and the sediment supply is illustrated in Figure 5.2.

Stream systems are dynamic in nature and, as a result, streamflow rarely is constant. During stormflow events, the rising limb of the hydrograph generally is related to high rates of sediment transport and degradation, as shown in Figure 5.3. However, as the peak streamflow discharge passes and the rate of streamflow discharge decreases, the amount of sediment in suspension also decreases and aggradation occurs.

a. Suspended load

Soil particles can be transported as a suspended load in the streamflow if their "settling velocity" is less than the "buoyant velocity" of the turbulent eddies and vortices of the water. The settling velocity is dependent largely upon the size and density of the soil particles, other factors being constant. The settling velocity of soil particles less than 0.1 mm in diameter generally is proportional to the square of the soil particle diameter, while the settling velocity of soil particles larger than 0.1 mm is proportional to the square root of the soil particle diameter.

Once the soil particles are in suspension, relatively little energy is required to transport the soil particles in the streamflow. In fact, a heavy suspended load will decrease the turbulence and make the transportation more efficient. Concentrations of suspended sediment are highest in streams where the depths are shallow and the velocities are high.

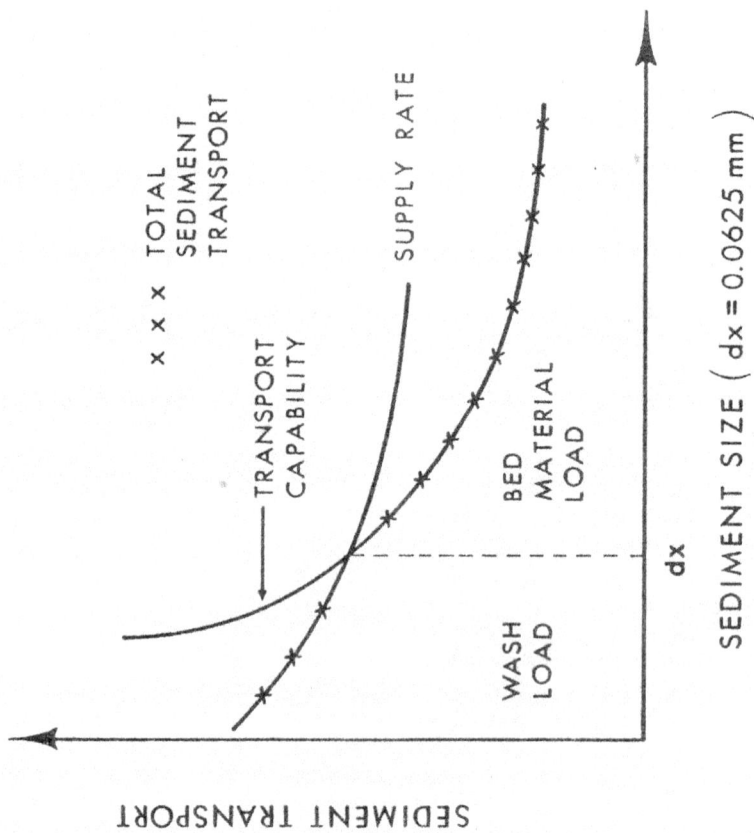

Figure 5.2. Rate of sedimentation as affected by the transport capability and supply rate for different size particles (Rosgen et al. 1980).

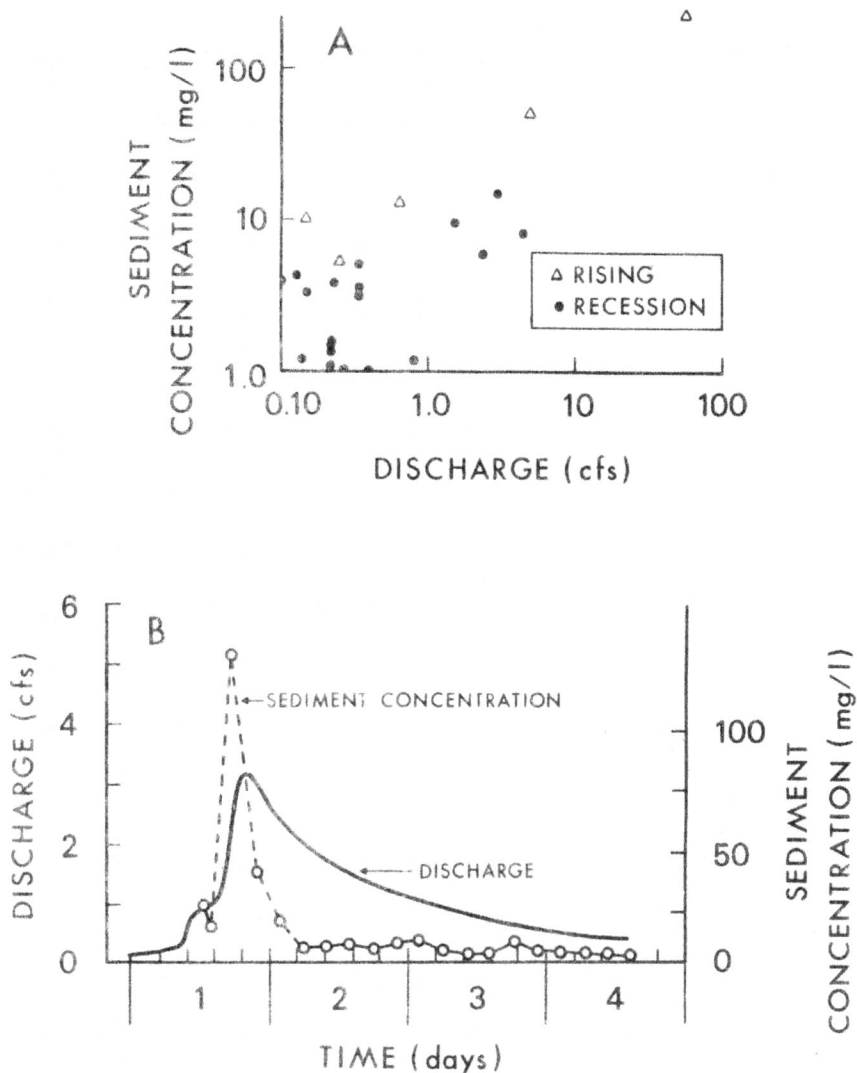

Figure 5.3. Example of a relationship between streamflow discharge and suspended sediment in the United States, with (A) representing the relationships for several streamflow events, including the streamflow event illustrated in (B) (Brooks et al. 1990).

The concentration of suspended sediment in a stream is lowest near the surface of the water, and increases with depth. Silt and clay particles less than 0.005 mm in diameter generally are dispersed uniformly throughout the depth. Larger soil particles are more concentrated near the bottom.

b. Bed load

Soil particles can be transported as bed load in groups or singly. These soil particles can be entrained if the vertical velocity of the eddies create sufficient suction to lift the soil particles from the bottom, or they can be started in motion if the force exerted by the water is greater on the top of the train than on the bottom. Soil particles can move by saltation if the hydrodynamic lift exceeds the weight of the soil particles. The soil particles will be re-deposited downstream if not re-entrained. Large, as well as small, soil particles can be rolled or slid along the bottom of the stream channel. The more rounded soil particles are moved more easily.

The large soil particles that a stream can move in traction as bed load is called the *stream competence*. The competency of a stream differs throughout the length of the stream and with time. Stream competency is increased greatly in periods of high peak flows and flooding events.

The force required to entrain a specific soil particle size is called the *critical tractive force*, and the velocity at which entrainment takes place is referred to as the *erosion velocity*. The DuBoys equation generally is used to calculate the critical tractive force as a function of the stream depth and channel gradient, and is written as:

$$T_f = W_w DS \tag{5.1}$$

where T_f = critical tractive force (kg/m^{2}0)
$\quad$ W_w = specific weight of water (kg/m^3)
$\quad$ D = stream depth (m)
$\quad$ S = stream channel gradient (m/m)

Equation (5.1) provides satisfactory estimates of the tractive force for streamflows of low velocities and small soil particle sizes, but it does not define the critical tractive force for streamflows of high velocities. For streamflows of high velocities and large soil particle size, the streamflow velocity generally is more important than streamflow depth and gradient, which has resulted in the *sixth-power law*, namely:

$$S_c = cV^6 \tag{5.2}$$

90

where S_c = stream competence

       c = constant

Another important concept in sediment movement is the *stream capacity*, which is the minimum amount of sediment of a specific soil particle size that a stream can carry in traction as bed load. An increase in the stream channel gradient and the streamflow discharge results in an increase in the stream capacity. If small soil particles are added to predominantly large soil particles in the stream bed, the stream capacity is increased. However, if large soil particles are added to small soil particles, the stream capacity is reduced. Stream capacity also decreases with increasing soil particle size.

All of the factors affecting the stream capacity are interrelated and vary with the stream channel geometry. Streams that carry large bed loads have shallow, rectangular or trapezoidal cross-sectional profiles, because there is a steep streamflow velocity gradient near the bed of these cross-sectional profiles. The typically parabolic cross sectional profiles of stream channels of watersheds in the ASEAN region do not have a steep streamflow velocity gradient near the stream bed.

5.1.3 Sediment Budgets

The transport or routing of sediment from the source areas on a watershed, where active erosion is taking place, to downstream river channels involves a number of complex processes. A *sediment budget* is a simplification of these processes and includes considerations of the:

- Rate of sediment movement from one temporary storage site to another.
- Amount of sediment and time in residence in each storage site.
- Linkages among the processes of transfer and storage sites.
- Changes in the character of the sediment as it moves through the system.

A sediment budget is a quantitative statement of the rates of production, transport, and discharge of the soil particles. Accounting for the variations in time and space of the transport and storage processes requires detailed models, a discussion of which is beyond the scope of this manual.

5.2 CLASSIFICATION OF SEDIMENT

Sediment can be classified in terms of its basic components, that is, suspended sediment and bed load. In addition, sediment can be classified in terms of soil particle size, for example, clay, silt, sand, gravel, cobble, and boulder. These size categories are:

	Size Range (mm)		
Clay	Smaller than 0.0039		
Silt	0.0039	-	0.0625
Sand	0.0625	-	2.0
Gravel	2.0	-	64.0
Cobble	64.0	-	256.0
Boulder	256.0	-	4096.0

The lithology of these soil particles is described by the usual rock or mineral names. Quartzite, sandstone, and basalt are examples of rock names, with quartz, feldspar, and magnetite as examples of mineral names.

5.3 UNITS OF MEASUREMENT

The concentration of sediment is expressed in terms of the weight of the sediment per unit volume of water-sediment mixtures, most commonly as:

- kilograms per cubic meter (kg/m^3).

Other units sometimes used to express the concentration of sediment include kilograms per tone (kg/t), milligrams per liter (mg/l), parts per million (ppm), and Newtons per cubic meter (N/m^3).

The weight of sediment per unit of time frequently is expressed as:

- kilograms per second (kg/s).

5.4 METHODS OF MEASUREMENT

Methods of measuring sediment involve the collection of representative samples, a laboratory analysis to determine the concentrations of sediment in the samples, and the appropriate data analysis.

Measurements of suspended sediment and bed load generally are made separately, because of the differences in soil particle sizes and in the distribution of the soil particles in a stream. Therefore, the methods of measurement also will be considered separately, as different techniques are employed in the measurement of each sediment component.

5.4.1 Suspended Sediment

Measurements of suspended sediment are based on sampling techniques, because measurements of "populations" are impossible in most instances. As suspended sediment concentrations vary with stream depth and the distance across a stream, a major problem in determining the suspended sediment amounts in a stream is obtaining a representative sample.

a. Sampling considerations

Sampling is the first of a series of steps that lead to the generation of data on the suspended sediment amounts in the streamflow. Care always must be taken to obtain a representative sample. If the sample is not representative, all of the care taken to provide an accurate analysis is lost. Importantly, the integrity of the sample must be maintained from the time of collection to the time of analysis.

The choice of the sampling locations and the sampling frequency will depend, in large part, upon the purpose of the investigation, the method of measurement, and the local conditions.

Sampling locations can be specified as:

- Individual locations in a network.
- A pair of locations above and below a site that is suspected of undergoing active geologic erosion or being subjected to improper land use practices.
- One location for sampling before and after an activity of potential sediment increase.

After the sampling locations have been selected, they should be referenced through detailed site descriptions. These descriptions, along with the method of measurement, should be used to identify all of the samples taken.

Sampling frequency is determined largely by the objectives of the investigation, the dynamics of the stream system in question, the accessibility of the sampling locations, and the personnel and facilities available in conducting the study. Sampling frequency in monitoring programs, which consists of repetitive, continuing measurements to identify variations and trends at the sampling locations, should include the collection of samples on a monthly basis, with special emphasis on extreme events.

b. Sampling methods

A number of methods are available to sample the suspended sediment load in a stream. Some of these methods also are used to collect samples for other water quality analyses (see Section 6.4.2 of this manual).

The collection of one or more "grab samples" is a common method, especially for relatively small streams. With this method, a container placed into a stream is oriented to allow the flowing water to enter the container. Once filled, the container is removed from the stream and referenced appropriately. However, this simple method may not be reliable because of the problem in obtaining a representative sample.

A single-stage sampler, consisting of a container with an inflow and outflow tube at the top, is illustrated in Figure 5.4. This type of sampler generally is used on small, fast-rising streams. A single-stage sampler begins its intake when the water-level exceeds the height of the lower inflow tube, and then continues until the container is full. Therefore, only the rising limb of a hydrograph is sampled, which can limit the usefulness of the data.

Depth-integrating samplers can minimize the sampling bias that is associated with a single-stage sampler. A depth-integrating sampler, such as the US DH-48, consists of an "aluminum fish" into which a container is placed, as shown in Figure 5.5. Immersed in a stream, water enters a nozzle, while displaced air is emitted through a vent. In operation, a depth-integrating sampler is lowered and raised at a constant rate in the stream. As a result, a relatively uniform sample of a vertical section of a stream is obtained. A number of these samples can be taken at selected intervals across the stream channel.

There has been increased effort to develop suspended sediment samplers which automatically pump a sample at pre-determined time intervals from a point in a stream. The advantage of such samplers is that they allow samples to be collected over a time period without personnel being present. The major components of an automatic sampler are the intakes, a pump, a splitter which draws off the desired volume of a sample, a circular table on which the containers for the samples are placed, and a water

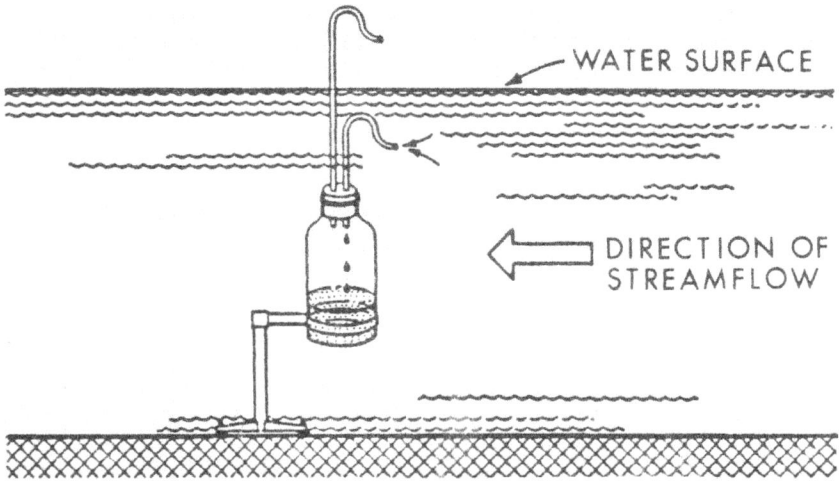

Figure 5.4. A single-stage suspended sediment sampler (Avery 1975).

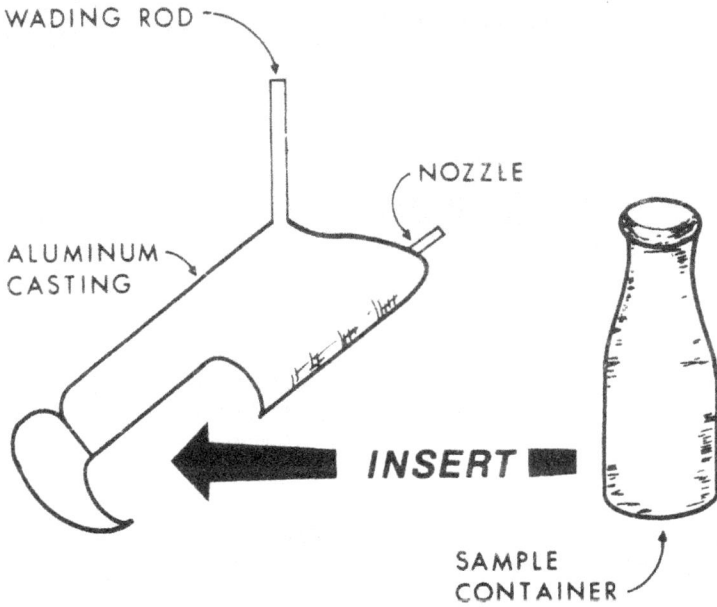

Figure 5.5. Diagram of a DH-48 depth-integrating suspended sediment sampler.

supply for priming the pump and flushing the sediment out of the intake before each sampling operation. In addition, there is a clock and control box with a sequence timer that initiates the required operations in the proper order. These components, with the exception of the intakes which are located at a point in the stream channel, are housed in a shelter, as shown in Figure 5.6.

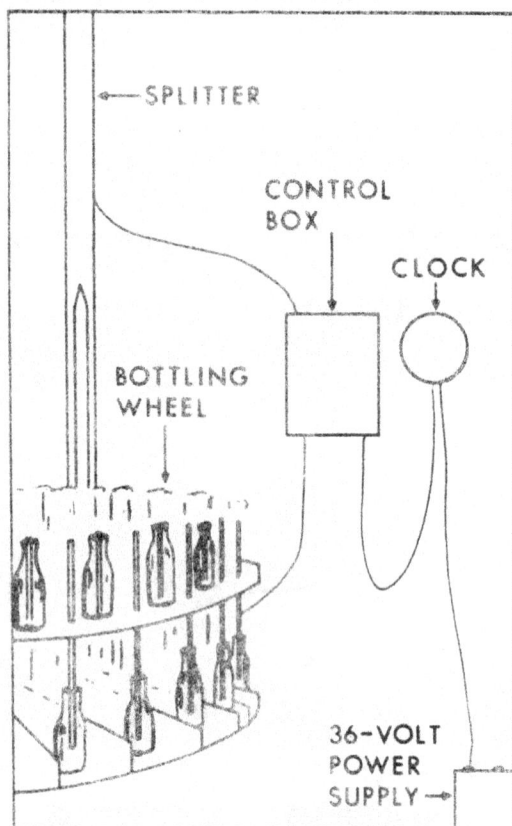

Figure 5.6. Component parts of an automatic suspended sediment sampler (Hansen 1966).

It generally is preferable to have an automatic sampler located adjacent to a continuous water-level recorder. An additional pen then can be installed in the recorder to make a "tick mark" on the chart when a sample is pumped. In doing so, a record of the water-level of the stream at which each sample is collected can be obtained. The stilling well provides a location for the floats that regulate the sampling frequency, or halt the sampling in the case of no streamflow.

Regardless of the method employed, each suspended sediment sample collected should be accompanied by a measurement of streamflow discharge to provide a complete record for subsequent analysis.

5.4.2 Bed Load

No one method of measuring the bed load materials is reliable, economical, and easy to use. Although a number of bed load samplers exist, few are utilized widely. Instead, estimates of bed load can be obtained through volumetric surveys and measurements of the weight per unit volume of the materials deposited on the watershed or in traps behind catchment basins. If the bed load deposits are relatively small, they can be weighed directly. In either case, these measurements subsequently can be partitioned into sands, gravels, and cobbles to determine the distributions by particle sizes.

The bed load materials that are transported by a stream also can be estimated from empirical formulas based on field and laboratory research data, although in many instances, these empirical formulas are site-specific.

5.4.3 Total Sediment

Measurement of total sediment is an essential part of evaluating watershed management practices. Such a measurement can be obtained from sediment measurement installations consisting of a catchment basin to trap the bed load and a series of splitters that collect a representative portion of the suspended sediment leaving the watershed.

A low dam is constructed to create a basin and provide the location for the splitters. Small streamflows that do not spill over the dam deposit their sediment in the basin. Some of the smaller sediment particles pass over the spillway during intermediate streamflows. During large streamflows, increasingly larger particles pass through the spillway. If the basin fills completely, the total sediment load will pass through the spillway. This situation must be prevented, however, because the larger sediment particles will not enter into the splitters. Therefore, the catchment must be large enough to prevent the basin from filling completely with sediment during ordinary streamflow events.

A continuous measurement of suspended sediment can be obtained with the series of splitters. As the water flows through the spillway, a small fraction is "split off," or separated, by a mechanical divider. This water is retained in larger containers for later analysis of the suspended sediment concentrations. By using a second and even a third splitter, relatively larger streamflows can be sampled by this method.

5.5   LABORATORY ANALYSIS

Laboratory analytical techniques for suspended sediment samples and bed load samples are described briefly in the following paragraphs.

5.5.1   Suspended Sediment Samples

Once a suspended sediment sample is obtained, the liquid portion of the sample is removed in a laboratory by filtering, evaporating, and when the equipment is available, centrifuging, and then the amount of sediment is weighed. The dry weight of the sample often is expressed as a concentration, as described in Section 5.6.1 of this manual.

a.   Filtration

To illustrate the laboratory analysis of a suspended sediment sample by filtration, a general procedure is described below:

- Weigh the filter paper.
- Filter an appropriate volume of the suspended sediment sample, for example, 100 cm^3.
- Dry the suspended sediment obtained through filtering at 103 degrees C. for a 24-hour period.
- Cool the sample and weigh the filter paper plus the suspended sediment in an air humidity controlled laboratory or desiccator unit.
- Subtract the weight of the filter paper from the weight of the filter paper plus the suspended sediment, the result being the weight of suspended sediment in the sample.

b.   Evaporation

A general procedure for the laboratory analysis of a suspended sediment sample by evaporation is:

98

- Weigh the container and the water-sediment mixture, and then subtract the weight of the empty container, the result representing the weight of the water-sediment mixture.
- Allow the water-sediment mixture to stand for a period of time for the settlement of the sediment from suspension.
- Decant the sediment-free water from the sample.
- Wash the remaining sediment from the container into an evaporation dish of known weight.
- Dry the sample at 103 degrees centigrade for a 24-hour period.
- Cool the evaporation dish in a desiccator cabinet.
  Weigh the evaporation dish and the contents, and then subtract the weight of the evaporation dish, the result representing the weight of suspended sediment in the sample.

c. Other methods

An alternative to the physical separation of the suspended sediment from the liquid portion of the sample is "turbidity" measurements, which can be used to index the suspended sediment load in a stream. Turbidity refers to the optical property of a sample that causes light to be scattered and absorbed rather than transmitted. The so-called Jackson Candle Turbidometer has been a standard instrument for turbidity measurements, with the results of these measurements reported in Jackson Turbidity Units (JTU). (One JTU is equal to 1 mg SiO_2/l.) However, due to the limitations of the Jackson Candle Turbidometer, photometric methods are becoming more common. These methods are adaptable to both field and laboratory conditions.

Correlations between turbidity measurements and suspended sediment can be developed for the stream being sampled. However, because turbidity measurements are based on optical properties of the sediment particles in suspension, attempts to extrapolate these correlations from one stream to another may not be successful.

5.5.2 Bed Load Samples

Samples of bed load materials can be analyzed in a laboratory according to the following general procedure:

- Weigh the empty container, after which a bed load sample is placed into the container.
- Evaporate the bed load sample to dryness.

- Transfer the bed load sample to an oven and dry at 103 degrees C. for a 24-hour period.
- Cool the sample and weigh the container plus the bed load sample.
- Subtract the weight of the empty container from the weight of the container plus the bed load sample, the result being the weight of bed load material in the sample.
- Partition the bed load material into sands, gravels, and cobbles, if desired, with a series of screens.

5.6   DATA ANALYSIS

5.6.1   Suspended Sediment Concentrations

After the weight of suspended sediment in a sample has been determined, it then can be expressed directly as a concentration in terms of kg/m^3, mg/l, ppm, and so on. Suspended sediment concentrations can be estimated by:

$$C = A \left[\frac{\text{weight of suspended sediment}}{\text{weight of water-sediment mixture}} \right] \qquad (5.3)$$

where C = suspended sediment concentration
A = factor which corrects for the differences in specific weights of water and water-sediment mixture, found in reference books on hydraulics and hydrology.

5.6.2   Bed Load

The amount of bed load materials in a sample generally is expressed in terms of total weight of the bed load, for example, in terms of t, kg, and so on. As previously mentioned, the total weight of the bed load can be partitioned into the weight of sands, gravels, or cobbles.

5.6.3   Total Sediment Load

To compute the total sediment load, the calculated weight of the suspended sediment simply is added to the weight of the bed load materials. Depending upon the purpose of the data analysis, total sediment loads for a stream can be computed for monthly, seasonal, or annual time periods.

5.6.4 Suspended Sediment-Streamflow Discharge Relationships

In many streams, there is a correlation between the suspended sediment load and the streamflow discharge. If sufficient data on suspended sediment and streamflow discharges are available, a relationship can be developed and used as a *sediment rating curve*. Such a relationship often takes a logarithmic form, as follows:

$$S = kq^m \qquad\qquad (5.4)$$

where S = daily sediment load (t)
$\quad$ q = daily streamflow discharge (m^3/s)
k and m = constants

5.6.5 Sediment Delivery Ratio

A commonly used method of relating erosion rates to sediment transport is the *sediment delivery ratio*, which is defined as:

$$D = \frac{Y}{T} \qquad\qquad (5.5)$$

where D = sediment delivery ratio
$\quad$ Y = sediment yield (weight/area/year)
$\quad$ T = erosion above the point at which (Y) is measured (weight/area/year)

The sediment delivery ratio of a watershed is affected by the climate, general physiographical position, texture of the eroded material, land use practices, and the local stream conditions. As the size of a watershed increases, the sediment delivery ratio generally decreases, a relationship that is illustrated in Figure 5.7.

5.6.6 Sediment Hydrograph

A *sediment hydrograph*, usually developed for suspended sediment, is a graphical representation of suspended sediment concentrations in relation to time. When superimposed on a streamflow hydrograph, the relationship between sediment transport and streamflow discharge becomes apparent, as illustrated in Figure 5.3.

It can be seen in Figure 5.3 that the higher rates of suspendend sediment concentrations generally correspond with the rising limb of the streamflow

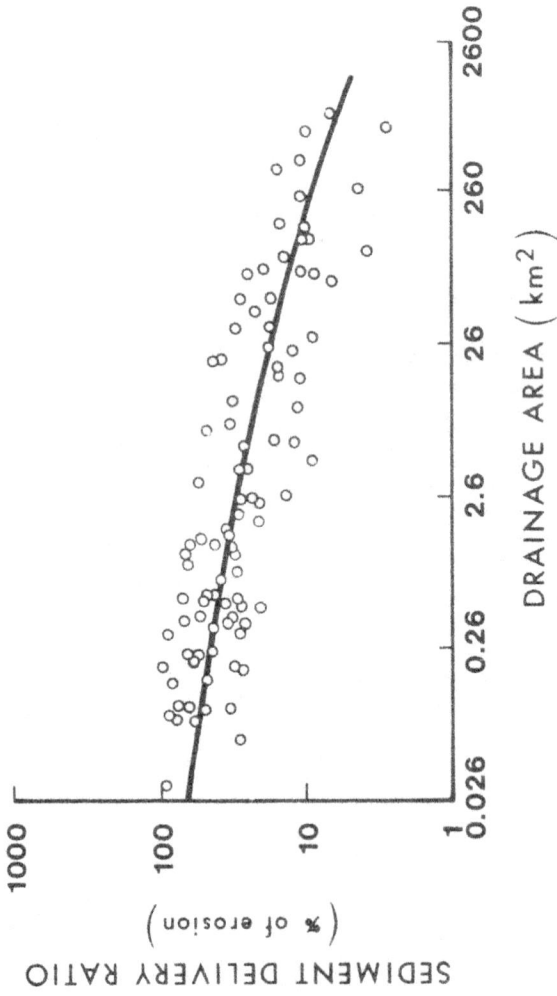

Figure 5.7. Sediment delivery ratio is determined from the size of a watershed (Roehl 1962).

hydrograph. However, once the flood peak passes and the streamflow discharge decreases, the suspendent sediment concentration also decreases, indicating that less sediment is available at the time of the stormflow event.

A sediment hydrograph, once developed for a specific stream, can be used to estimate the sediment yield from a watershed, assuming that suspended sediment concentrations are related to streamflow discharge measurements. In essence, calculations of streamflow discharge then can be used as "predictors" of suspended sediment concentrations. These predictions of suspended sediment, in turn, are "integrated" through time for the duration of the streamflow event in question. To obtain an estimate of total sediment yield, estimates of bed load also must be available.

CHAPTER SIX

Water Quality

6.1 WATER QUALITY CHARACTERISTICS

The criteria for water quality depends upon the intended application of the water. Principle categories include agricultural use, commercial use, human consumption, and water as part of the natural environment as a habitat for fish and other aquatic lifeforms.

Data which expresses the condition of water can be either numeric or narrative. Numeric criteria are specific for the application of the water. Narrative critera are more general and open to interpretation with respect to the intended application.

The qualities of water may be divided as

- Physical characteristics
- Dissolved chemical constituents
- Bacteriological quality

6.1.1 Physical Characteristics

Physical characteristics include the temperature, turbidity, color, and suspended solids (undissolved matter.) In addition, the physical characteristics may be extended to include the movement, direction of flow, and rate of evaporation, depending on the nature of the water quality analysis report.

The temperature of water is of paramount importance as a biological habitat for fish and other aquatic lifeforms, but seldom directly considered in any other application, such as agricultural or commercial use. However, the temperature of the water has other indirect effects, such as the saturation level of dissolved gases. These gases include beneficial oxygen, as well as gases that are toxic. So the overall relationship between temperature and the health of the water is more complex than one might imagine.

Many species of fish have a distinct preference for the temperature of water, and a variation of only a few degrees will greatly influence reproduction. Changes in the temperature from season to season must also be considered when appraising the temperature of waters as a habitat. The annual average temperature is not sufficient to provide a meaningful analysis in any case. The temperature at various depths should also be considered.

Turbidity of water may be due to sediment or microorganisms including algae. Cloudy water reduces the amount of light that can penetrate the surface, affecting aquatic plants rooted in the bottom. Sediment also adversely affects many species of fish and interferes with their reproductive capability.

The color of water is affected by both dissolved ions as well as undissolved solids and microscopic lifeforms, such as algae. Algae is usually seen as a yellow to green hue in water, but some species can produce hues of red and other colors. Soil is generally gray to brown in color. Some mineral deposits can produce strong coloration, notably iron oxides (reddish brown) and some copper salts (bluish-green.) Complex organic substances in water can produce almost any color, but are most frequently seen as a translucent milky white discoloration. Natural water appears to be blue when it is very clear and sufficiently deep, because it absorbs more of the light at the red end of the spectrum and reflects slightly more of the blue wavelengths. This is in the category of an optical illusion, and not actual coloring of the water.

Suspended solids in water are most often soil, including clay and sand. Solids may also include microscopic plants, and waste of any type (including human and animal waste, industrial waste, etc.) Suspended solids directly affect water clarity, and thus reduce the efficiency of photosynthesis, but they may also bind to toxic compounds including heavy metal ions, and effectively increase the concentration of contaminants in a local region as a result.

106

6.1.2 Dissolved Chemical Constituents

This category of water quality includes the following:
- Dissolved inorganic ions
- pH
- Dissolved gases
- Dissolved organic components
- Electrical conductivity (which relates to all of the above)

Some inorganic dissolved ions are essential for aquatic life, but the same ions may be harmful or lethal in higher concentrations. The ions which are of the greatest concern are mercury and certain other metals. These metals are harmful to both the fish themselves, and to the food supply when the fish are caught and consumed later.

Electrical conductivity is a direct measure of the concentration of ions in water but the numerical data says nothing about which specific ions are present. Conductivity is a product of all ions present, including dissolved gases that dissociate. The temperature that the conductivity measurement is made is also a factor in the numerical result obtainted. To know the chemical makeup of the water, separate tests must be performed for each ion, including those normally present even in drinking water, such as sodium, calcium, nitrogen, bicarbonate, and chlorine.

The pH value of water is a measure of the hydrogen ion content, measured on a logarithmic scale between 0-14, with 7 being neutral. Most aquatic lifeforms can exist between pH 5.5 and 8. It is normal for the pH to vary slightly across a body of water, particularly near the shoreline when there is runoff from soil, or any kind of contamination entering the watershed.

Many gases can be found in small concentrations in water. The amount of dissolved oxygen (DO) will directly affect biodiversity. Some fish require high concentrations of dissolved oxygen, while other species (such as carp and catfish) actually prefer low concentrations of oxygen. When the water contains 90% or more of the saturation level of dissolved oxygen, it is considered to be a healthy environment for aquatic lifeforms, including plants and microscopic organisms. The saturation level is temperature-dependent. Cold water can contain more dissolved gases (including oxygen) than warm water. Polluted water has a lower saturation level.

Most of the dissolved oxygen in water is as a result of the interaction of the surface with the atmosphere, and this will also vary with season, partially from rainfall and also from wind that mixes the surface with the air.

Some dissolved gases are highly toxic, notably ammonia, chlorine and hydrogen sulfide. Ammonia is naturally occurring in low concentrations, but all three of these gases are common by-products of industry. Even trace amounts of chlorine or hydrogen sulfide can destroy vast regions of aquatic organisms. Ammonia levels are increased by man from the run-off of fertilizer on farms near the watershed, and also from dumping of untreated sewage. The amount of ammonia present in the water will vary with temperature (warm waters contain less dissolved gases) and other contaminants present in the water (some of which will react with ammonia.) All gases are relatively easy to remove. Simply boiling water will significantly reduce the concentration of dissolved gases. Other contaminants are just as harmful to life, and much more difficult to reduce to a safe level once introduced into a watershed

One of the most important forms of contamination of water for all purposes is the presence of heavy metal ions, most notably mercury, cadmium and lead. These ions are destructive to aquatic habitats. They contaminate the food supply, both from the fish that are caught in polluted waters, as well as through the plants irrigated with the same water. The plants grown with this water may be used as feed for livestock, thereby also contaminating meat supplies. These ions can leach into underground water tables and ultimately find their way into wells used for drinking water, and also as water for livestock.

Heavy metal ions accumulate in the body. One of the ways in which they cause direct harm is by binding to the metal ion site in metalloenzymes. Because they bind more strongly than the metal that is usually present (such as magnesium or zinc) once they are introduced, they can not be removed. The metalloenzyme either functions incorrectly, or doesn't function at all as a result. The consequences of this lead to numerous ailments and diseases when enough metalloenzymes are taken over. These ions are purged from the biological system very slowly, generally lasting for many years, so accumulation over time is facilitated. Frequent exposure results in a host of maladies.

Mercury is especially problematic in water supplies. First, because the toxic levels are so small that they are difficult to detect, and can reach the watershed in several ways.

The inorganic mercury which is often present in trace amounts is usually not the problem itself. The problem is that it is converted to methylmercury at the bottom of bodies of water when the conditions are right. These conditions include an acidic pH level, organic matter, and low levels of dissolved oxygen.

Dissolved organic components in water can be either harmful or beneficial. There are essential nutrients in any aquatic ecosystem that are necessary for a healthy environment. These are sources of food for the microscopic members of the food chain, without which the entire ecosystem can not continue. When the concentration of nutrient materials are very much greater than the natural "background level" they are considered detrimental to the aquatic habitat.

One of the most toxic manmade substances contaminating watersheds is the class of chemical called polychlorinated biphenyls, or PCB's. These industrial compounds were banned in the United States in 1977, but continue to leach into the environment almost everywhere in the world. They are toxic, and almost certainly carcinogenic.

Another important organic source of pollution are pharmaceuticals, and in particular, birth control hormones which interfere in the reproductive activity of fish and amphibians. Trace amounts in sewage and other human contamination have impacted many aquatic environments.

Toxic substances are defined as any material or the decomposition product of any material, upon which exposure, ingestion or inhalation will cause illness, cancer, genetic mutation or interfere with reproduction of the organism exposed, or the offspring of the exposed organism. As a general rule, any agent which causes these problems in an aquatic life form, eventually has similarly detrimental effects on the human population in the surrounding area.

6.1.3 Bacteriological Quality

Total coliform bacteria are a collection of relatively harmless microorganisms that live in the intestines of man and animals. They aid in the digestion of food. A specific subgroup of this collection is the fecal coliform bacteria, the most common member being Escherichia coli. These are distinguished by their ability to grow at elevated temperatures and originate from the fecal material of warm-blooded animals. The source can be untreated sewage, runoff from land in which livestock are grazing, pets, or even naturally present wildlife including aquatic birds.

Total coliform (TC) bacteria, fecal coliform (FC) bacteria, and fecal streptococcus (FS) bacteria often are related to the physical characteristics of the streams. Coliform bacteria counts frequently appear to be dependent upon the flushing effects of surface runoff from intensive rainfall events.

In some situations, accumulated of organic material on the ground surface can become a bacteria filter. It has been found, for example, that runoff that has percolated through a strip layer of organic material often contains fewer bacterial counts than runoff that has not passed through the strip.

A major source of fluctuations in coliform bacterial counts, especially FC bacteria, that can lead to pollution problems is a concentration of warm-blooded animals (including native wildlife species, livestock, and humans) on many watersheds, especially in areas that are adjacent to the streamflow networks. Therefore, to keep the FC bacteria counts below pollution levels, land management practices that reduce the concentrations of all sources of potential bacterial contamination need to be encouraged.

6.2 WATER QUALITY MEASURES

6.2.1 Water Quality Criteria

Water quality criteria specify the concentrations of water quality constituents which, when adhered to, are expected to result in aquatic ecosystems suitable for the "ultimate uses" of water. These criteria, usually expressed in terms of numerical limits, reflect the current knowledge on the effects of surface water and groundwater pollutants on health and welfare. Water quality criteria are derived from scientific facts that are obtained from experiments or field observations of the behavior of the water quality constituents' response to an introduced meterial or stimulus in regulated environmental conditions for specified time periods.

Water quality criteria are not necessarily intended to offer the same degree of safety for the survival and propagation for all organisms at all times within a specified aquatic ecosystem. However, these criteria are designated to protect the essential life in the water and the direct users of this life and the water.

An ideal data base for the development of water quality criteria consists of information on a large proportion of aquatic species in a particular ecosystem. Furthermore, this information should show the response of these aquatic species to a range of concentrations for a tested constituent for a long period of time. This information generally is not available for all situations, although investigators are beginning to obtain this information for a few of the water quality constituents.

Some of the water quality constituents often have small criterion levels for human health considerations. In routine water quality sampling, only a few laboratories are capable of measuring these constituents to this level of accuracy. These low concentrations indicate high toxicity and potential health problems, suggesting that any amount found in a water sample requires additional testing.

6.2.2 Water Quality Standards

The word *criterion* should not be used interchangeably with the word *standard*, in considering water quality measures. A "criterion," as described above, represents a water quality constituent concentration that is associated with a degree of environmental effect, upon which a scientific judgement is based. It has come to be a designated concentration of a constituent that will protect a specific organism, a community of organisms, or a prescribed water use. A "standard" represents a legal entity for a particular section of a streamflow network.

A water quality standard may use a water quality criterion as a basis for regulation or enforcement. A water quality standard may differ from a water quality criterion, however, because of prevailing local conditions, such as naturally occurring organic acids, or because of the importance of the particular section of a streamflow network, economic considerations, or the desired degree of safety to a specific aquatic ecosystem.

Water quality standards generally are rules that determine the use or uses to be made of a water body and the water quality criteria necessary to protect that use or uses. These standards are to be enforceable and are developed through a process that includes social, legal, economic, and institutional considerations. Water quality standards also serve as a basis for water quality-based treatment controls.

6.2.3 Water Quality Index

An index is a number, usually dimensionless, whose value is considered to express a measure of the relative magnitude of a condition. A water quality index is one that can be used in the planning and operation of a water resource project, and for the dissemination of information to the public.

The setting of a water quality index depends largely upon the knowledge and experience of the developer, that is, the rating and weighting of a chosen water quality constituent is totally at the disposal of the developer. A particular water quality index cannot be compared with another water quality index, and does not necessarily indicate the level to which the water should be treated economically.

A water quality index can be expressed as a vector, as follows:

$$WQI = f[x_1, x_2, ..., x_n] \tag{6.1}$$

where WQI = water quality index
x_1 = total organic carbon
x_2 = suspended solids
x_3 = coliforms, and so on

It often is possible to reduce "n" by using a computer program to calculate a correlation coefficient matrix. From this correlation coefficient matrix, those water quality constituents not correlated significantly with others can be omitted. It also might be possible to reduce "n" through consultation with the published literature. The number of water quality constituents may be reduced, for example, by using past experience in addition to a knowledge of the source of pollutants and the designated water use.

6.3   WATER QUALITY MONITORING

To evaluate the kinds and amounts of water quality constituents in water, representative water quality measurements are taken. In most situations, a sampling procedure is required, as continuous measurements of water quality are difficult to obtain and costly. Therefore, a water monitoring plan is necessary. Such a plan should be based on knowledge of the water system being sampled, an understanding of the time and space distribution patterns of the water quality constituents to be sampled, and most importantly, the purpose for which the water quality monitoring is being done.

6.3.1   Purposes of Monitoring Programs

Water quality monitoring programs are established to answer specific questions about water quality and, therefore, should be designed accordingly. Water quality monitoring programs in watershed management generally can be classified as one of the following:

- *Cause and effect monitoring* - monitoring which is conducted to determine the effects of specific actions on water quality constituents. For example, a water quality monitoring program may be established to determine the effects of timber harvesting on suspended sediment concentrations.

- *Baseline monitoring* - monitoring undertaken to help a watershed manager determine if there are specific trends in water quality constituents at a particular location, and if there are, whether the trends are changing in time.

- *Compliance monitoring* - monitoring conducted to determine if specified water quality standards are being adhered to. An example is a monitoring program designed to determine whether the streamflow at a particular location is suitable for drinking water.

- *Inventory monitoring* - monitoring is designed to indicate the existing water quality conditions. From the results' of the monitoring program, sites that are suitable for specific developments can be selected.

6.3.2 Design Criteria for Monitoring Programs

Regardless of the purpose of the water quality monitoring program, the general design of the program incorporates the following procedure:

- Define the objectives of the monitoring program. Identify the potential sources of pollution and other needs for monitoring.
- Choose the sampling locations. To the extent possible, attempt to link the sampling locations to sites for other monitoring operations.
- Select the water quality constituents to sample.
- Determine the sampling frequencies.
- Decide upon the methods required to sample the selected water quality constituents. Select the equipment necessary for field analyses and, if needed, a suitable laboratory.
- Calculate the costs for the monitoring program, including the initial capital expenditures and recurring sampling and analytical costs.
- Determine the methods of data analysis and the system of data storage and subsequent retrieval to be used. Preliminary sampling can help in selecting the appropriate analytical techniques.
- Decide upon the reports that will be prepared and when they will be presented. These reports should meet the objectives of the monitoring and the informational needs of watershed management.

6.4 METHODS OF MEASUREMENT

Methods of measuring water quality require the collection of representative samples, an analysis to determine the concentrations of the water quality constituents, and the appropriate data analysis.

The water quality constituents considered in this section of the manual are those commonly considered in analyzing the impacts that can result from forestry, agricultural, and grazing activities on upstream watersheds. Most of these pollutants originate from non-point sources, and then enter streams from land areas in a diffused manner.

6.4.1 Sampling Considerations

Choice of the sampling locations, selection of the water quality constituents to sample, and determination of the sampling frequencies are all important in the planning of water quality monitoring programs.

a. Choice of sampling locations

Sampling locations are chosen to meet the monitoring objectives. Sampling locations can be established as:

- Individual locations in a network.
- A pair of locations above and below a suspected impact.
- One location for sampling before and after an activity of potential impact.

A network of locations frequently is used in water quality monitoring programs. The locations can be concentrated in key sites, or the locations can be distributed throughout the entire watershed, as illustrated in Figure 6.3. It often is useful to select locations that are easily accessible and can be reached throughout the monitoring period. However, in doing so, it is important not to preclude the selection of locations that will contribute to a more representative and complete data base.

Detailed site descriptions should be made of the sampling locations, once these sampling locations have been chosen. These site descriptions then should be used to identify all of the samples taken.

114

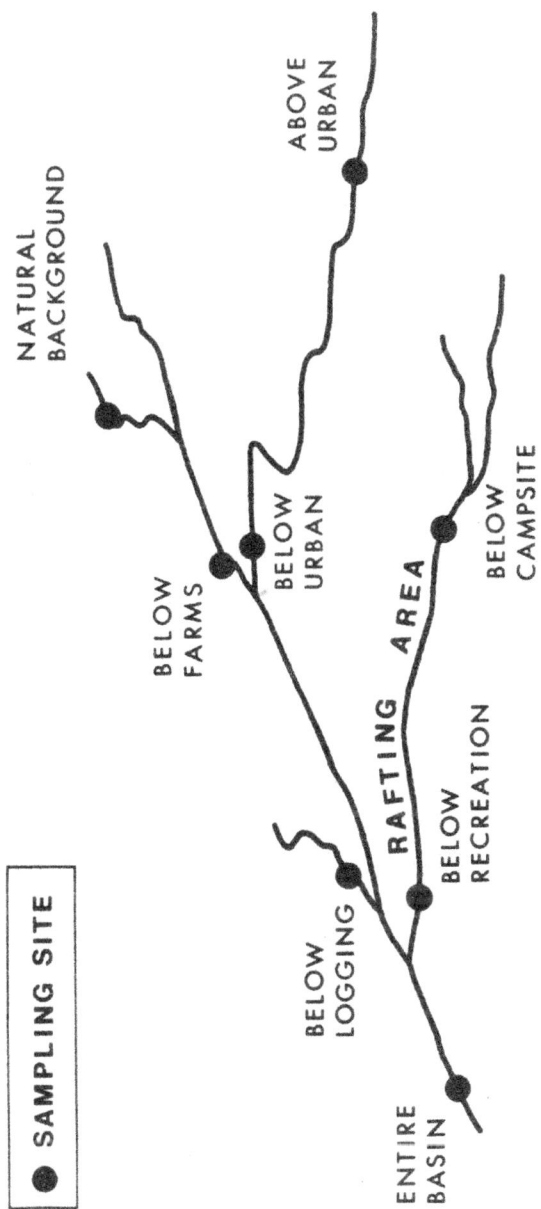

Figure 6.3. Example of a network of locations used in a water quality monitoring program (Kunkle et al. 1987).

115

b. Selection of water quality constituents

Selection of the water quality constituents to sample is dependent, in large part, upon the objectives of the monitoring program. The rationale for the selection of specific water quality constituents in analyzing the impacts of forestry, agricultural, and grazing activities on water quality is presented in Table 6.1.

Table 6.1. Rationale for the selection of water quality constituents (Kunkle et al. 1987).

Constituent	Rationale
Biochemical oxygen demand (BOD)	Index of potential lowering of dissolved oxygen levels caused by the decomposition of organic wastes
Chemical oxygen demand (COD)	A quick estimate of BOD
Chloride	Indicator of animal wastes and sewage
Conductivity	Indicator of total dissolved solids and an index of inorganic pollutants
Dissolved oxygen (DO)	Indicator of lowered oxygen levels caused by organic sediment loading
Fecal coliform (FC) bacteria	Indicator of contaminants caused by animal and human wastes
Fecal streptococcal bacteria	FC:FS ratio can indicate origin of fecal (FS) contamination
Herbicides	Analyze for locally-applied herbicides
Nitrate-nitrogen	Indicator of nitrogen-containing fertilizer or pesticidal contamination
Phosphate	Indicator of phosphate-containing fertilizer contamination

Table 6.1. continued.

Constituent	Rationale
pH	A valuable measure for interpreting solubility ranges of dissolved chemical constituents
Temperature	A requirement for pH and conductivity, and a valuable measure for interpreting solubility ranges of dissolved chemical constituents
Total coliform (TC) bacteria	Indicator of fecal contamination, although contains more soil organisms and other naturally occurring bacteria and, therefore, not always an indicator of pollution
Total dissolved solids (TDS)	Index of inorganic pollution
Total suspended solids (TSS)	Indicator of sediment transport
Turbidity	Estimator of TSS

c. Determination of sampling frequencies

Determining the sampling frequencies for the water quality constituent is a combination of art and science, considering both the practical constraints to sampling and the field conditions. In this manual, an arbitrary distinction is made among *key water quality constituents, supplementary water quality constituents*, and *special studies*. Key water quality constituents, in the opinion of the watershed manager, are the major determinants of the quality of streamflow in a specific area. Supplementary water quality consituents, also important in characterizing the quality of streamflow, occur either in lesser concentrations than the key constituents or less regularly in the streamflow. Special studies require the sampling of water quality constituents normally not included in a

117

monitoring program, but thought to be potentially important intermittently in terms of water quality analysis.

Key water quality constituents are sampled more frequently than are supplementary water quality constituents, although both should be sampled on a systematic time schedule. Special studies are made as required to sample water quality parameters in the appropriate time or on a schedule that will provide the "best" results. Suggested sampling frequencies for key and supplementary water quality constituents are listed in Table 6.2.

In addition to the sampling frequencies shown in Table 6.2, extra sampling of specific water quality constituents can be important in periods of high streamflow, low streamflow, and stormflow.

Table 6.2. Sampling frequencies for key and supplementary water quality constituents (Kunkle et al. 1987).

Stream	Constituent	
	Key	Supplementary
Perennial	Sample every two weeks in principal impact periods. Sample monthly in remainder of year	Sample quarterly in conjunction with key constituents
Intermittent	Sample every two weeks in principal impact periods. Sample monthly in remainder of year	Sample quarterly in conjunction with key constituents
Ephemeral	Sample twice a year in periods of rainfall events resulting in streamflow	Sample once a year in conjunction with key constituents

Due to the diluting effects of high streamflows and the pollutants that are introduced by surface runoff, high streamflow periods can be critical times to sample. Water quality consituents best analyzed in these periods include:

- Sediment and turbidity originating from erosion of roads, timber harvesting areas, agricultural lands, and other areas lacking a vegetative ground cover.

- Constituents that are adsorbed into sediment particles and subsequently are transported off a watershed.

Other water quality constituents are best detected in periods of low streamflow, when these constituents become concentrated as a result of the small volume of water. A major source of streamflow in the periods of low streamflow often is groundwater inflow from waters that have had the longest residence-time to contact mineral solids and mobilize the constituents. Among the water quality constituents best sampled in low streamflows are:

- Metals, organics, and chloride.
- DO minimum values, especially in situations where low streamflow and elevated temperature coincide.
- BOD and COD maximum values, which generally occur with DO minimum values.

Stormflow events, similar to periods of high streamflows, result in surface runoff that carry contaminants into stream networks, the difference being stormflows can occur at any time of the year. Water quality constituents that are transported primarily in stormflow periods include:

- Sediment from timber harvesting operations.
- Fecal bacteria from areas of livestock grazing, nitrogen and phosphorus from fertilizers, and pesticide residues.

d. Statistical considerations

It frequently is assumed that the water quality constituents occur in a random manner in streamflow. This being the case, it is possible to estimate the number of samples required to estimate the mean concentrations of these constituents, within specified confidence intervals, from equation (2.1), presented in section 2.3.1 of this manual.

6.4.2   Sampling Methods

Some of the methods commonly used to sample water quality in streamflow, many of which also are used to sample the suspended sediment load in a stream, are described briefly below.

Much of the sampling in water quality monitoring programs is accomplished with grab samples. Often, a grab sample obtained in a clean glass or plastic container is satisfactory. However, a grab sample is representative of the streamflow discharge only at the time of sampling. Some water quality constituents are affected by the magnitude of the streamflow discharge. Streamflow discharge measurements, therefore, should accompany all water quality samples.

Depth-integrating samplers and samplers which automatically pump a sample also are used to sample for water quality. Descriptions of these samplers, which also are used for suspended sediment concentrations, are described in section 5.4.1 of this manual. Automatic samplers are activated either to pump a sample at a specified time interval or when the stage of the stream attains a specified level.

For some of the more common water quality constituents, instrumentation has been developed to facilitate the "direct sampling" of a constituent in the aquatic environment. Standardized oxygen and pH probes, for example, are available to sample the dissolved oxygen and hydrogen ion concentrations in a streamflow system. Similarly, electrode probes can be used to determine the concentrations of many dissolved chemical constituents in water.

It generally is recommended that all water quality samples be collected by a depth-integrating sampler, whenever it is practical to do so. An exception may be in the case of shallow streams, where the depth is insufficient to allow "true" depth integration. In these instances, grab samples that are collected at one or more verticals across the stream channel are appropriate. However, the container must be held carefully just beneath the water surface to avoid disturbing the stream bed.

6.5   HANDLING AND TREATMENT OF SAMPLES

It is not only necessary to ensure that the field samples obtained are homogeneous, but also that they are homogeneously sampled in the analysis. Therefore, for the analysis, the sample should be well stirred and all particles should be kept in suspension.

120

6.5.1 Preservation

Deteriorated water samples negate all of the efforts and costs in obtaining a "good" sample. The shorter the elapsed time between the collection of a sample and its sample analysis, the more reliable will be the analytical results. Some water quality constituents must be analyzed in the field to obtain reliable results, because the composition of the sample almost always will change before it arrives at a laboratory. However, some water quality samples can be preserved satisfactorily by chilling, by adding a suitable acid or germicide, or by other special methods of preservation.

Determinations of temperature, pH, specific conductance, and dissolved gases (for example, dissolved oxygen and carbon dioxide) should be made in the field. Water quality samples for metal analysis can be preserved by the addition of nitric acid. Samples for determination of organic constituents are preserved by chilling or freezing. Samples for the determination of pH, nitrate-nitrogen, phosphates, and sulfactants are preserved by chilling the sample immediately in an ice bath and storing the samples in the dark at a temperature slightly above freezing. Some preservatives interfere with the methods of laboratory analysis and, therefore, must be selected carefully.

Methods of preservation are relatively limited and are intended to delay biological action, delay the hydrolysis of chemical compounds and complexes, reduce the volatility of the water quality constituents, and reduce absorption effects. Techniques for preservation generally are limited to pH control, chemical addition, refrigeration, and freezing.

Complete preservation of water quality samples is a practical impossibility. Regardless of the nature of the sample, complete stability for every water quality constituent is never achieved. Preservation only can delay the chemical and biological changes that inevitably will continue after the sample is removed from a stream.

6.5.2 Filtration

It often is necessary to filter the water quality samples for an analysis, for example, of dissolved chemical constituents. When this is the case, the appropriate pore-diameter membrane (0.45 um) should be used. It generally is advisable to discard the first 150 to 200 ml of a sample to rinse the filter and filtration apparatus of any contaminating substances. This technique minimizes the risk of alteration of the composition of the sample in the filtering operation.

Other filtering procedures, including the designation of the type of filter to use, are specified in references on the analysis of other water quality constituents.

6.5.3 Storage

Water quality samples collected for metal analysis that have been preserved with nitric acid can be stored for several months. If no sediment is present, chilled or refrigerated samples also generally are stable. However, most water quality samples should be analyzed as soon as possible, within the time limitations specified by the analytical method.

A good practice is to always store water quality samples in a cool place, out of the sunlight. A container that is compatible with the analysis for the specified constituents should be used for the storage of all samples.

Glass and polyethylene containers, for example, are suitable for storage in most cases. However, polyethylene containers are not recommended to store samples for color analysis. Polypropylene caps generally are recommended whenever the sample acidity can cause metallic caps to corrode.

6.6 WATER QUALITY ANALYSIS

Water quality analysis can take place in the field or laboratory, either by *photometric* or *titrametric methods*, as discussed in the following paragraphs.

6.6.1 Photometric Methods

Photometric methods relate the intensity of a color to the concentration of the water quality constituent present, following the addition of a specified amount of reagent to a standard volume of the water sample. Instruments required consist of a light source, a photocell, and a light meter. Interposed in the light beam are a color filter and the sample, as shown in Figure 6.4. The amount of light that is absorbed is proportional to the concentration of the constituent in the sample. Photometric methods of analysis are adaptable to field use.

6.6.2 Titrametric Methods

Titrametric methods involve the slow addition of reagents of known concentrations to a standard sample of water. An abrupt color change

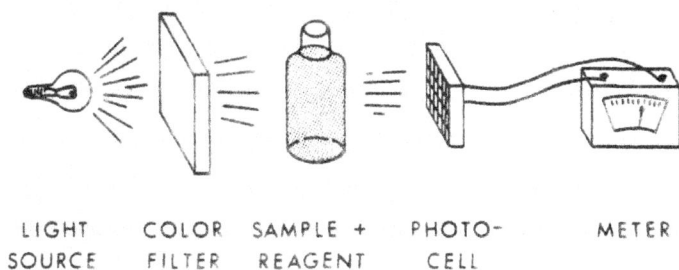

LIGHT   COLOR   SAMPLE +   PHOTO-   METER
SOURCE   FILTER   REAGENT   CELL

Figure 6.4. General photometric method for determining the quality of water (Avery 1975).

occurs after a certain amount of reagent is added to the sample, as illustrated in Figure 6.5. The concentration of the water quality constituent of concern is proportional to the amount of reagent added to the sample to produce the color change. Titrametric methods are better suited to laboratory conditions and, in general, provide more accurate results than the photometric methods.

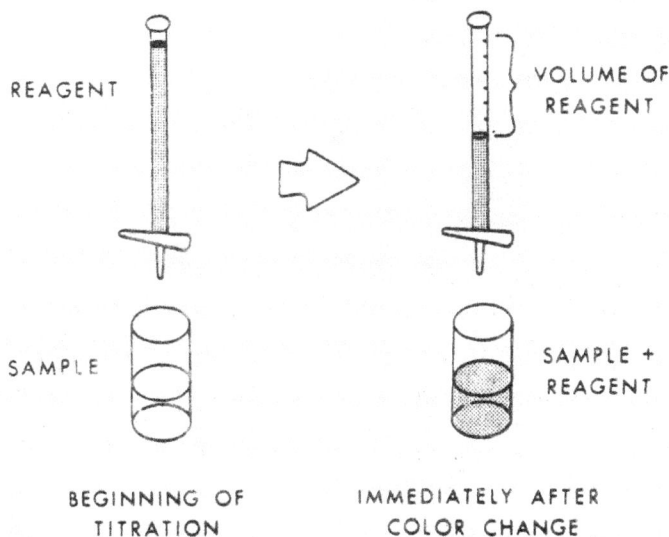

REAGENT   VOLUME OF
                                           REAGENT

SAMPLE   SAMPLE +
                                           REAGENT

BEGINNING OF   IMMEDIATELY AFTER
TITRATION   COLOR CHANGE

Figure 6.5. General titrametric method for determining the quality of water (Avery 1975).

123

6.7 LABORATORY ANALYSIS METHODS

Laboratory analysis methods for many of the commonly encountered water quality constituents are outlined below. In cases where more than one laboratory technique is available for a constituent, which frequently is the case, the method outlined generally is considered to require a minimum of laboratory facilities, and apparatus and materials cost. For the water constituents not considered, references on the appropriate laboratory analytical techniques should be consulted.

Preparation of reagents and standard chemical solutions, a requirement for the analysis of specific water quality constituents, is prescribed in references on the laboratory techniques for the analysis of water quality constituents and, therefore, will not be discussed in detail in this manual.

6.7.1 Conductivity

A conductivity meter (for example, the HACH Conductivity Meter Model 2511, as utilized in the procedure described below) is required for this laboratory analysis of water. The general procedure for analyzing the conductivity of the water sample is:

- Turn the instrument ON.
- Set the selecting switch to position 5.
- Immerse the probe into the water quality sample.
- Observe the reading of the probe.
- Select the appropriate range for the selecting switch.
- Record the reading.

In some situations, the conductivity of the sample can exceed the range of the instrument, in which the procedure is:

- Dilute the water sample with de-ionized water to a final sample volume of 100 cm^3.
- Measure the conductivity as outlined above.
- The conductivity of the sample is calculated as:

$$C = \frac{100 \, [c_i - c_d(100 - V_s)]}{V_s} \qquad (6.2)$$

where C = conductivity

C_i = conductivity reading on instrument

C_d = conductivity of diluted sample

V_s = volume of sample

6.7.2 Dissolved Oxygen (DO)

Either chemical or instrumental methods can be employed to analyze the DO content in a water sample. A chemical method is outlined below. In applying this chemical method, it is necessary to prepare a number of reagents, including a manganese sulfate solution, an alkali-iodide-azide reagent, concentrated sulfuric acid, a starch solution, a sodium thiosulfate stock solution, a standard (0.0250 N) sodium thiosulfate titrate solution, and a special potassium fluoride solution. The laboratory procedure is:

- To a water sample in a 250 to 300 ml bottle, add 2 ml of manganese ulfate solution, followed by 2 ml of the alkali-iodide-azide reagent.
- Using a stopper, cap the bottle carefully to exclude air bubbles and mix the test solution by inverting the bottle 15 times.
- When the precipitate settles, leaving a clear supernate, shake the bottle once again.
- Allow a 2 minute period of contact between the precipitate and salt water.
- After at least a 2 minute settling has produced 100 ml of clear supernate, carefully remove the stopper and immediately add 2 ml of concentrated sulfuric acid by allowing the acid to flow down the neck of the bottle.
- Re-cap the bottle with the stopper, and mix the test solution by inverting the bottle until the dissolution is complete.
- Distribute the iodine uniformly throughout the bottle before the decanting of the sample volume needed for titration. (Use a sample volume corresponding to 100 ml of the original water sample after a correction for the loss of the sample through displacement with the reagents.)
- Titrate the sample with the standard thiosulfate titrate solution to a pale "straw" color.
- Add 1 to 2 ml of the starch solution and continue the titration to the initial disappearance of the blue color.
- If the "end point" is overrun, back-titrate with 0.0250 N biniodate solution added dropwise or adding a measured volume of the sample. (Correct for the volume of the biniodate solution or added sample. Disregard subsequent recolorations.)

- In this laboratory procedure, for 200 ml of the original water sample, 1 ml of (0.0250 N) sodium thiosulfate stock solution = 1 mg/1 DO. To obtain the results of the laboratory analysis in mm of oxygen gas per 1 unit of the sample, corrected to 0 degrees centigrade and 760 mm of atmospheric pressure, it is necesary to multiply the calculated mg/1 DO value, obtained from a solution of either equation (6.3) or (6.4), by 0.70. Toexpress the results in terms of percent saturation at 760 mm of atmospheric pressure, the appropriate solubility data and equations for correcting the solubilities to barometric pressure values other than that of mean sea level (found in references on the laboratory analysis of water quality constituents) must be used.
- Calculate the solubility of oxygen for temperature values between 0 and 30 degrees C. by:

$$mg/1 \ DO = \frac{0.678[P - u]}{35 + t} \qquad (6.3)$$

where mg/1 DO = solubility of oxygen in distilled water
P = barometric pressure (mm Hg)
u = saturated vapor pressure (mm Hg)
t = temperature

- For temperature values between 30 and 50 degrees C., calculate the solubility of oxygen from:

$$mg/1 \ DO = \frac{0.827[P - u]}{49 + t} \qquad (6.4)$$

6.7.3 Fecal Coliform (FC) Bacteria

Materials needed for the laboratory analysis of FC bacteria in a water quality sample described below are:

- Millipore sampler for coliform or Coli-count sampler
- Incubator

The procedure for the analysis of FC in the sample is:

- Remove the stick (with the grid lines) from the clear case, allowing nothing to touch the grid to protect the sterilization.

- In distilled water, the sampler can be dipped directly into the water sample. Alternatively, fill the clear case to the upper line with a sample of the water to be tested. (It is important that the water should not be contaminated by bacteria from fingers.)
- Insert the sampler into the filled case, shake several times, and then hold the case still for 30 seconds. (During this time, the sampler will absorb 1 ml of water.)
- Remove the sampler and shake it several times to remove excess water.
- Empty the case and shake it to remove excess water.
- Insert the sampler into the empty case.
- Place the entire apparatus into an environment that maintains a temperature of 35 degrees C., and leave in-place for 18 to 24 hours.
- After 18 to 24 hours, examine the paper grid for signs of FC bacterial colonies. (There can be many spots on the sampler, but only those which are raised, shiny, and either blue or blue-green are colonies from a single FC bacteria.)
- Count the number of colonies.

The presence of FC bacteria indicates that pathogenic organisms may be present. More than two colonies on the sampler generally means that the water is unsafe for drinking. However, the presence of FC bacteria is only an indicator of contamination. Because its detection involves incubation and sterile conditions, the laboratory analysis of FC bacteria is more difficult to do correctly than many water quality tests described in this manual.

Negative results should not necessarily be accepted without repeating the laboratory analysis. Caution should be exercised in interpreting the results to note whether a recently occurring rainfall event could have increased the contamination.

6.7.4 Nitrate-Nitrogen

The following apparatus are required in the laboratory analysis of nitrate-nitrogen in a water sample:

- Nitrate ion-selective electrode
- Digital pH meter
- Magnetic stirrer bar
- Reference electrode

Preparation of reagents, specifically distilled water and a standard nitrate solution, is necessary. The procedure for the analysis of nitrate-nitrogen involves the construction of a calibration graph, as described below:

- Add by pipette 25 ml of the standard nitrate solution to a 100 ml beaker containing the magnetic stirrer bar.
- Remove the nitrate electrode from the solution in which it has been immersed and place it in a holder with the reference electrode.
- Rinse both electrodes with de-ionized water, remove the surplus water with tissue paper, and immerse the electrodes in the test solution.
- Note the depth of the immersion and keep this level constant throughout the analysis.
- Stir the test solution at a "moderate" rate, keeping the stirring constant throughout the procedure.
- When the electrodes are reading a steady value, record the electromotive force (E_1) value.
- Repeat the above steps by using at least four other standard nitrate solutions, spanning the range of 0.4 to 400 ppm nitrate or 0.1 to 150 ppm nitrate-nitrogen (E_2, E_3, E_4,...E_n).
- Plot the emf values against the concentrations of nitrate-nitrogen on semi-logarithmic paper.
- Calculate the calibration slope (k), which should be approximately - 56 mV for a ten-fold increase in concentration of nitrate-nitrogen from:

$$k = \frac{E_1 - E_2}{\log s_1 - \log s_2} \qquad (6.5)$$

Once the calibration graph has been constructed:

- Add by pipette 25 ml of test solution to a 100 ml beaker containing the magnetic stirrer bar.
- Repeat the above steps, recording the steadying values E_x.
- Read the concentration of nitrate-nitrogen in the test solution from the calibration graph.

6.7.5 Orthophosphate

In the laboratory analysis of the orthophosphate in a water sample, a Spectronik Spectrophotometer is required. In preparing the reagents for this analysis, a molybdate and sulfite reducing reagent is necessary. The analysis procedure is:

- Filter the water sample, if it is not clear.
- Transfer a suitable portion to a 100 ml volumetric flask and dilute to 80 ml.
- While swirling the sample in the flask, add 5 ml of molybdate reagent, followed by 5 ml of sulfite reducing reagent.
- Dilute to the mark and mix the solution thoroughly.
- Measure the absorbance of the solution.
- Determine the orthophosphate (OP_{mg}) in the solution from the prepared calibration curve.
- Calculate the orthophosphate (OP_{ppm}) in the solution by:

$$OP_{ppm} = \frac{OP_{mg}[10^6]}{V_s[10^3]} \qquad (6.6)$$

6.7.6 Percent Light Transmittance

A colorimeter (for example, the HACH AC-DR Colorimeter, as used in this method) is required for this laboratory analysis, the procedure of which is outlined as follows:

- Turn the instrument ON 10 to 15 minutes before the analysis.
- Insert the percent transmittance meter scale into the colorimeter, using the 4,000-5,000 color filter with distilled water in the sample cell.
- Adjust the colorimeter to a reading of 100 percent transmittance.
- Pour a water sample into the sample cell and place into the light cell.
- Observe and record the percent transmittance reading on the meter scale.

6.7.7 Pesticides

For the purposes of laboratory analysis, pesticides often are classified as chlorinated hydrocarbons, organophosphoric compounds, or other compounds. The first two classes of pesticides can be analyzed by gas-liquid chromatography, because they include atoms that are measured by element-selective detectors. These electron-capture and microcoulombic-detectors are "highly sensitive," particularly to the chhlorinated pesticides.

A program of pesticidal analysis is a specialized topic, the details of which are presented in references on the analysis of water quality constituents.

6.7.8 pH

A pH meter (for example, the Model PHM 62) is used in this laboratory analysis. The procedure for the analysis is:

- Prepare the electrodes for the measurement of pH.
- Make the following setting on the pH meter:

  - pH-MV button released.
  - HOLD button released.
  - TEMP knob at the actual measuring temperature.

- Adjust against the pH buffer:

  - Dip the electrodes into a beaker containing the buffer solution and stir.
  - Turn the BUFFER knob until the meter reading agrees with the known pH of the buffer solution.

- Rinse the electrodes off with distilled water.
- Immerse the electrodes into the sample being tested and stir.
- Observe and record the pH of the sample.
- Remove the electrodes, rinse off with distilled water, and leave in the distilled water.

6.7.9 Settleable Solids

The laboratory procedure for analyzing the settleable solids in a water sample, which requires the use of an Imhoff cone, is presented as:

- Fill the Imhoff cone to the liter-mark with a thoroughly mixed water sample.
- Allow the sample to settle for 45 minutes, then gently stir the sides of the cone and allow the sample to settle for 15 minutes.
- Observe and record the volume of settleable solids in the cone..

6.7.10 Temperature

The procedure is:

- Immerse a thermometer into the water sample.
- Wait until equilibrium is reached.
- Observe and record the reading on the thermometer.

6.7.11 Total Hardness

The hardness of water generally is due to the dissolved calcium and magnesium salts in the sample. The total hardness of water, using an Solochrome Black (Eriochrome Black T) Indicator, can be analyzed by complexometric titration, as described below:

- Add 4 cm^3 of Guffer solution to a 50 cm^3 sample of the water to be tested, followed by 30 mg of hydroxylammonium chloride, 50 mg of AR potassium cyanide, and 30 to 40 mg Solochrome Black Indicator mixture.
- Titrate the solution with standard EDTA solution (0.01 M) until the color is changed from red to pure blue.
- If there is no magnesium present in the solution, it is necessary to add 0.1 cm^3 of magnesium-standard EDTA solution before adding the indicator.
- Express the total hardness of the water in parts of $CaCO_3$ per million parts of water.

6.7.12 Turbidity

A turbidimeter (in this case, a HACH Turbidimeter Model 2100 A) is necessary for this laboratory analysis. The general procedure is:

- Standardize the instrument by selecting the desired range, placing the appropriate standard into the cell holder, covering with the light shield, and adjusting the STANDARDIZE control to obtain a meter reading equal to the FTU value of the standard.
- When standardizing the instrument in the 100 and 1,000 FTU range, the cell riser must be used.
- Pour 25 cm^3 of the water sample into a sample cell.
- Place the sample cell into the cell holder and cover with the light shield.
- Observe and record the FTU value of the water sample.

6.7.13   Total Solids (TS) and Total Suspended Solids (TSS)

The laboratory analysis for TS and TSS in water samples requires the following apparatus:

- Evaporating dish
- Muffle furnace
- Stream bath
- Analytical balance
- Desiccator
- Drying oven
- Suction flask
- Filtering apparatus
- Glass fiber fitter disks without organic binders

The procedure for determining the TS in a water sample is:

- Weigh the dish used for evaporation.
- Add 20 ml of the water sample.
- Evaporate to dryness.
- Transfer the sample to the oven and dry for 30 minutes at 103 degrees C.
- Cool the sample.
- Weigh the sample on the analytical balance.
- Calculate the total solids by:

$$TS = \frac{1000[A - B]}{V_s} \qquad (6.7)$$

where A = weight of dish plus sample
       B = weight of dish

The procedure for determining TSS is:

- Weigh the filter paper.
- Filter a suitable volume (100 cm^3) of the water sample.
- Transfer the sample to the oven and dry for 1 hour at 103 degrees C.
- Cool the sample.
- Weigh the sample on the analytical balance.
- Calculate the total suspended solids:

$$TSS = \frac{1000[A - B]}{V_s} \qquad (6.8)$$

6.7.14 Total Dissolved Solids (TDS)

From equations (6.7) and (6.8), the total dissolved solids in a water sample are calculated as:

$$TDS = TS + TSS \qquad (6.9)$$

where TDS = total dissolved solids

6.8   DATA ANALYSIS

6.8.1   Concentrations of Water Quality Constituents

Information on water quality most commonly is presented in terms of the concentrations of the water quality constituents sampled. To conform with the present laboratory analytical techniques and the current procedures in most of the laboratories analyzing the water quality constituent, concentrations of water quality constituents often are specified in terms of mg/l (milligrams per liter) and ug/l (micrograms per liter), rather than in ppm (parts per million) and ppb (parts per billion). All of these expressions of the concentrations of water quality constituents are useful in characterizing the quality of water samples.

It generally is advisable to be consistent in the use of these terms in reporting the results obtained from water quality monitoring programs. Concentrations of specific water quality constituents, for example, should be reported in the consistent terms to provide a basis for comparisons and evaluations through time.

The proper number of significant figures to be used in reporting the concentrations of the water quality constituents is specified to indicate the confidence that the data user can expect. Importantly, for all reported water quality data sets, there is expected to be an uncertainty of one unit in the last significant figure.

The standard deviation for a determination of concentration of water quality constituent indicates the probable uncertainty of the measurement in a statistical framework. Therefore, this measure of statistical variability should be calculated and presented in reporting the water quality characteristics of streamflow.

6.8.2 Water Quality Measures

Water quality measures, including water quality criteria, water quality standards, and water quality indices, are developed, in large part, from baseline information on the concentrations of water quality constituents in the streamflow.

CHAPTER SEVEN

Climate

The term *climate* generally is defined as "the average conditions" of weather patterns, over a period of years, as measured by rainfall, air temperature, relative humidity, evaporation, wind, and solar radiation and radiant energy. *Weather*, in turn, is a set of atmospheric conditions for a specified time and, therefore, refers to events. Climate, which is basic to watershed management, is difficult to describe because of the frequent deficiencies in the length and consistency of necessary meteorological records.

Climate is characterized most easily from the long-term records of the World Meteorological Organization of the United Nations or from data sets that are collected from instruments and equipment at local weather stations. When the information describing the climate of a region is obtained from a local weather station, it is important to maintain long-term, high-quality data collections. To achieve this goal, it is necessary to specify the:

- Site requirements for a weather station.
- Specific measurements to be taken.
- Instruments and equipment to be used in the data collections.
- Operating instructions and maintenance for the instruments selected.

7.1 SITE REQUIREMENTS FOR A WEATHER STATION

7.1.1 Location

A weather station should be located in large openings, away from obstructions and sources of surface moisture and dust particles. A weather station also should be installed on level ground, where there is only a low vegetative cover. Furthermore, the weather station should be situated to receive full sunlight for the greatest possible number of hours.

The following "guidelines" should be considered in the selection of a site for the location of a weather station:

- Locate the weather station in a place that is representative of the conditions in the area of concern. Consider the vegetative cover type, topographic features, elevation, and local weather patterns.
- Select a site that will provide for long-term operation and relatively unchanged exposures. Consider the developmental plans for the area (including plans for roads and buildings), ultimate obscurations through the growth of vegetation, and the observers availability.
- Arrange the weather station to provide data that are representative of the specific area in which the weather station is located. Consider the exposure requirements for the instruments and equipment in relation to prevailing wind patterns, the movements of the sun,topography, vegetative cover, nearby reflecting surfaces, and obstructions to the wind.

The following situations should be avoided when selecting a site for the location of a weather station:

- Sources of surface moisture, such as irrigated areas, lakes, swamps, and rivers. If unavoidable, locate the weather station a minimum of 100 m to the windward side of the source.

- Sources of dust particles, for example, roads. If unavoidable, once again, locate the weather station a minimum of 100 m to the windward side of the source.
- Large reflecting surfaces, for example, white-painted buildings. If unavoidable, locate the weather station far enough away from the reflecting surface to prevent artificial shading, at least a distance equal to the height of the reflecting surface or 15 m, whichever is the greatest.

- Large buildings, trees, and dense vegetation. Locate the weather station at least a distance equal to the height of the obstruction.
- Changes in topography, such as gullies, peaks, ridges, steep slopes, and narrow valleys.

7.1.2 Layout

The instruments and equipment for a weather station should be arranged to allow a free flow of air and full exposure to the available sunlight. Ideally, the weather station should be large enough to accommodate additional instruments and equipment, as these become necessary. Graveled paths to the instruments and equipment are desirable, otherwise the paths will become muddy in rainy weather and dusty in dry weather. A ground-plan for a weather station is presented in Figure 7.1.

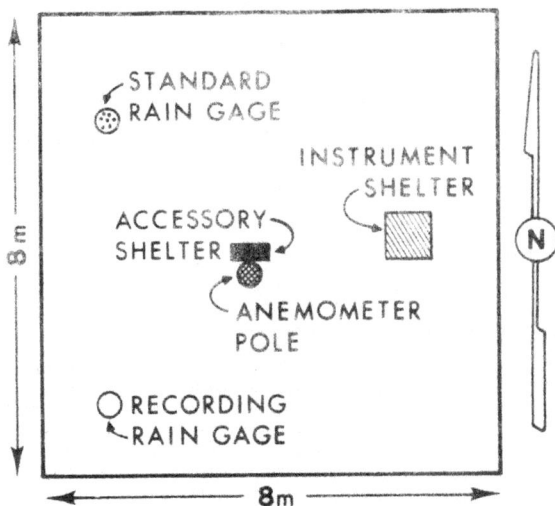

Figure 7.1. Ground-plan for a weather station (Fischer and Hardy 1972).

A fence around the weather station is not essential, unless there is a danger of the instruments being upset or otherwise damaged by curious people or animals. However, a fence often improves the appearance of a weather station and tends to discourage tampering by unauthorized personnel.

7.2   AIR TEMPERATURE

Simply stated, *air temperature* is a measure of the degree of "hotness" or "coldness" of the air. Air temperature measurements taken routinely at a weather station are maximum temperature and minimum temperature obtained from a maximum-minimum thermometer. A continuous record of air temperatures are obtained from a thermograph.

7.2.1   Maximum-Minimum Thermometer

A maximum-minimum thermometer, actually two thermometers, consists of a mercury and a spirit thermometer fixed in a spinning bracket. When exposed properly in an instrument shelter, these two thermometers become an accurate means of measuring the maximum and minimum air temperatures at a weather station.

a.   Instrument

A maximum thermometer is mercury-filled and has a small constriction in the fine bore of the tube above the bulb. As the mercury in the bulb expands with increasing air temperature, some of the mercury in the column is forced past this constriction. When the air temperature drops, the mercury above the constriction can not retreat into the bulb. When the bulb-end of the maximum thermometer is lowered to a reading position, therefore, the top of the mercury column indicates the highest air temperature for the period of measurement. After being read, a maximum thermometer is reset by whirling in its mount. Through this action, the mercury is forced through the constriction and back into the bulb.

A minimum thermometer, which is alcohol-filled, has a small glass index rod floating in the bore. This index moves freely in the alcohol. When the air temperature drops, the alcohol contracts and retreats down the bore. As the alcohol column contracts, it "drags" the index with it, by means of the surface tension at the top of the column. When the air temperature again rises, the alcohol flows past the index, leaving it positioned at the lowest air temperature attained. A minimum thermometer is reset by turning it upside down in its mount until the index returns to the top of the alcohol column.

b.   Installation

A maximum-minimum thermometer equipped with an appropriate support is designed to be exposed in an instrument shelter. The support is fastened

138

to the center of shelter cross-board, making sure that the support is oriented so that the spinning clamp is on the bottom.

The maximum thermometer is mounted in the lower spinning clamp, two-thirds of the way up the stem from the bulb, as shown in Figure 7.2. If mounted too near the middle, the mercury column can separate when whirled and part of it can become lodged at the top of the stem. The thumbscrew on the clamp is tightened to hold the thermometer in place when it is whirled.

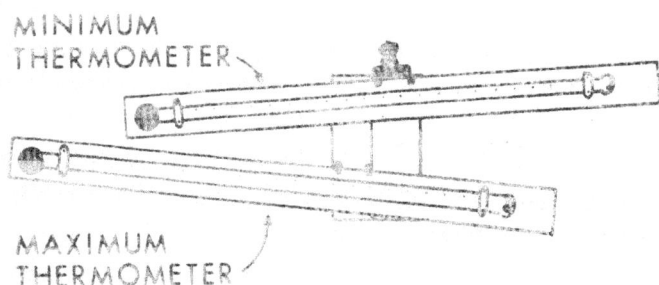

Figure 7.2. Maximum-minimum thermometer mounted on a support. The maximum thermometer is in the lower or spinning bracket. The minimum thermometer is in the upper bracket (Fischer and Hardy 1972).

When properly mounted, the bulb end of the maximum thermometer is on the left and raised about 5 degrees above the horizontal. This position facilitates the flow of mercury as the air temperature rises, and minimizes the chance of the mercury retreating through the constriction as the air temperature drops.

The minimum thermometer is mounted in the upper spinning clamp, slightly above the middle of the stem (Figure 7.2). The thumbscrew is tightened on the clamp to hold the thermometer in place.

When properly mounted, the bulb end of the minimum thermometer is on the left and approximately 5 degrees below the horizontal. This position facilitates the downward movement of the index when the air temperature drops, and minimizes the chance of the index rising as the air temperature rises. This position also prevents the accumulation of vaporized liquid above the column or resulting bubble formation.

c. Operating instructions

The following precautions should be observed when reading any type of liquid-in-glass thermometer:

- Do not touch the glass or place hand near the bulb.
- Do not breathe directly onto the thermometer. Keep face back, except when making the final reading.
- If the instrument is hand-held, stand in the shade. Whenever possible, face into the wind.
- Avoid parallax errors when reading the thermometer. The observer's eye should be positioned such that a straight line from the eye to the meniscus or the index forms a right-angle with the thermometer stem.
- Check the reading before recording.

Instructions for obtaining the "correct" readings and settings of a maximum-minimum thermometer include:

- Read the minimum thermometer first.

    - Read the minimum temperature from the upper end of the index.
    - Read the current temperature from the top of the alcohol column.
    - Do not reset the thermometer at this time.

- Read the maximum thermometer.
    - Unlock and slowly lower the thermometer to a vertical position, so that the mercury column is resting on the constriction in the tube.
    - Read the maximum temperature from the top of the mercury column.

- Reset the maximum thermometer first.

    - Whirl the thermometer until the reading agrees (within 1 degree C) with the reading at the top of the alcohol column of the minimum thermometer.
    - Lock the thermometer to a nearly horizontal position.

    - Reset the minimum thermometer last.

140

- Invert the thermometer until the index drops to the end of the alcohol column.
- Return the thermometer to a nearly horizontal position.

d. Maintenance

Thermometer maintenance generally is concerned with cleaning, restoring obscured markings, and recognizing and correcting defects such as broken constrictions and separated mercury or alcohol columns.

Annual maintenance requirements for a maximum-minimum thermometer include:

- After removing the thermometers from the mounting plates, use soap and water to clean the thermometers and mounting plates. Remove dirt and corrosion with cleaning solvent or vinegar.
- Carefully check the thermometers for defects. Repair or replace as necessary.
- If the markings on the thermometers are obscured, renew the markings by spreading a small amount of lampblack oil color on the stem and then immediately rub off the excess with a piece of hard-finish paper.

Periodic maintenance requirements are:

- Add oil to the spinner, as needed.
- Check both thermometers for defects.
- Check the thumbscrews on the clamps for tightness.
- Dust the thermometers with a soft brush to remove any accumulated dirt.

7.2.2 Thermograph

A continuous record of air temperatures frequently is an essential requirement in describing the climate of a region. The durations of different temperature levels often are significant in watershed management. Thermograms obtained from an instrument called a thermograph provide the only adequate means of measuring air temperatures on a continuous basis. This instrument also is invaluable in obtaining a measure of air temperature at weather stations that can not be visited more often than once a week.

a. Instrument

A commonly used form of a thermograph has a sensitive unit consisting of a flattened, curved, liquid-filled bourdon tube or a bimetallic element, depending upon the model of the instrument. Changes in the air temperature alter the curvature of the tube or the element. One end of the tube or the element is fixed, while the other end actuates a pen mechanism that moves across a strip-chart on a drum that is motivated by a clock movement.

The time-scale of a thermograph usually is 8-day-per-revolution of the clock drum. Pen readings on a thermogram generally are checked weekly against a maximum-minimum thermometer at the same location, so that the degree of accuracy can be noted and the appropriate correction applied to the data taken from the thermogram.

A theromgraph often is combined with a relative humidity recorder into one instrument, referred to as a *hygrothermograph*. With a hygrothermograph, air temperatures and relative humidity are recorded simultaneously on a double-scaled strip chart.

b. Installation

A thermograph usually is installed on a shelf or on the floor of an instrument shelter. In situations where it is not possible to install the instrument in a shelter, a thermograph with a "thermoshield" should be considered to reduce the radiant heat effects.

c. Operating instructions

At each visitation to a weather station where a thermograph is located, the instructions generally followed are:

- Remove the chart.
- Record any correctional data on the chart.
- Place a "new" chart on the drum, after writing the actual date on the chart.
- Wind the clock.
- If necessary, fill the pen with ink.
- Place the drum in place.
- Rotate the drum by hand until the pen is positioned on the correct line.

- If necessary, re-adjust the pen to indicate a "more correct" air temperature.

d. Maintenance

The maintenance requirements of a thermograph differ with the type, make, and model of the instrument. Therefore, the manual furnished by the manufacturer of the thermograph should be consulted for the specific maintenance details.

7.2.3 Data Collection and Processing

Hourly air temperature values, when required in watershed management planning, are obtained through interpretations of thermograms. In addition to these values, the mean daily air temperature is the most commonly required piece of information about the air temperature at a weather station. Mean daily air temperature values are obtained from data collected either by amaximum-minimum thermometer or a thermogram.

a. Data collection

Air temperature information obtained from a maximum-minimum thermometer usually is collected on a daily basis, although a longer time period can be monitored when required. Continuous readings of air temperatures on a thermogram generally are obtained on a weekly basis. Chart replacements on the thermograph usually are made at this approximate time interval.
The collection and reduction of data from a maximum-minimum thermometer are straightforward. However, the collection and reduction of data from a thermogram requires more detailed procedures.

b. Data processing

The processing of data from a thermogram includes chart annotations, tabulation of the data, and calculations. The chart annotations on a thermogram require the observer to:

- Enter the location of the instrument and the name of the observer.
- Enter the chart number and the dates of record.
- Enter the dates and watch times of chart placement, inspections, and removal.

143

- Note the date on the centerline of each day of record to facilitate tabulations.
- Enter the air temperature readings from a maximum-minimum thermometer or the dry bulb of a psychrometer at the time of the removal of the chart as checks on the pen readings.
- Enter notes to show any malfunctioning of the instrument or clock.

The tabulation of the data from a thermogram to calculate the mean daily air temperature involves one step:

- Enter the daily maximum and minimum air temperature values taken from the thermogram. These respective values normally are recorded for the 24-hour time period immediately proceeding the 8 AM localtime.

The mean daily air temperature is calculated from:

$$T_{mean} = \frac{T_{max} + T_{min}}{2} \qquad (7.1)$$

where T_{mean} = mean daily air temperature (degrees C)
T_{max} = daily maximum temperature (degrees C)
T_{min} = daily minimum temperature (degrees C)

The "daily air temperature" that often is reported is in reality a "median air temperature," based upon daily readings of the maximum and minimum air temperatures. However, as approximations of the "true" mean air temperature, this procedure frequently is erroneous, because the median value generally is higher than the mean value.

Once a long-term record of mean daily air temperatures at a weather station is available, the mean monthly air temperature, which is the average of all of the mean daily temperatures for each month, and the mean annual air temperature, which is the average of all of the mean daily air temperatures for the year can be determined.

7.3 RELATIVE HUMIDITY

Relative humidity is a ratio, expressed in terms of percent, of the actual amount of water vapor in the air compared to the total amount of water vapor that is necessary for saturation. The dew point temperature is the temperature at which the air, if cooled, will attain saturation. At this

144

temperature, the rela ive humidity is 100 percent and dew will start to form on a surface. instruments that measure relative humidity are a psychrometer, wet- and dry-bulb thermometers, a hygrometer, and a hygrothermograph.

7.3.1 Psychrometer

Several types of psychrometers are used to measure the relative humidity, including the electric fan psychrometer, the hand fan psychrometer, and the sling psychrometer. These psychrometers differ mainly in the method of ventilating the thermometers. The sling psychrometer is described in the following paragraphs.

a. Instrument

A sling psychrometer consists of two carefully matched mercury thermometers with cylindrical bulbs, placed side-by-side on a common mounting plate, as shown in Figure 7.3. A thin cotton wick covers the bulb of one of the thermometers. When this wick is wet and air passes the thermometer bulbs, evaporation from the "wet" bulb causes its temperature to drop. Assuming that adequate ventilation is present, the amount of evaporation from the wick depends upon the moisture content of the surrounding air mass.

The lower the relative humidity of the surrounding air, the greater the evaporation from the wet bulb wick and, therefore, the greater spread between the wet bulb temperature and that of the dry bulb. The dry bulb temperature indicates that of the surrounding air. Wet and dry bulb temperatures obtained are used to determine the relative humidity percent and the dew point temperature from standard psychrometric tables.

b. Operating instructions

A sling psychrometer is ventilated by whirling the instrument in a circular pattern around the head of the observer. To obtain relatively good accuracy, the observer should stand in a shaded open area, away from buildings and obstructions that may be struck during whirling.

If the light conditions permit, the observer should face into the wind to avoid any influence of body heat on the thermometers. Importantly, do not let sunlight or rain strike the instruments.

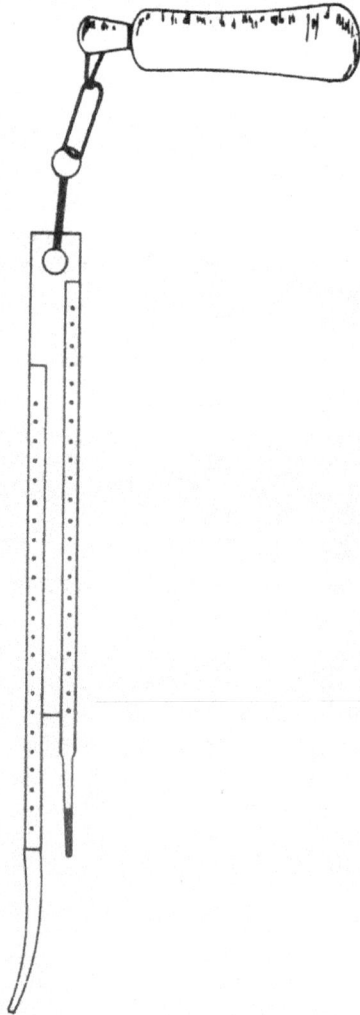

Figure 7.3. A sling psychrometer (Fischer and Hardy 1972).

Specific operating instructions for a sling psychrometer are:

- Check the instrument, making sure that the handle and chain are in good condition. Check the columns of the thermometers for separation. Be sure that the thermometers are mounted firmly to the housing.
- Check the wick to be sure that it is clean, of the proper length, and securely fastened to the bulb.
- Saturate the wick with mineral-free water of air temperature.
- Ventilate the thermometers by whirling the instrument at a full arm's length, with the arm parallel to the ground, for 1 minute at a rate of approximately 200 whirls per minute.
- If the relative humidity is low and shade is not available, premature drying of the wet bulb can be a problem. To prevent this problem, whirl the sling psychrometer in body shade, open to breezes for a few minutes before whirling in sunlight if possible. This action often will reduce the amount of whirling required.
- Note the wet bulb temperature, then whirl another 40 to 50 times and read the temperature again. If the wet bulb temperature is lower than the first reading, continue to whirl and read until the temperature will not decrease. When the lowest temperature is reached, read and record the temperature.
- Read the dry bulb temperature immediately after the lowest wet bulb temperature is obtained.

c. Maintenance

Annual maintenance of a sling psychrometer includes:

- Remove the retaining clips and lift the thermometers from the mounting plate. Remove the wick and discard. Maintain the thermometers in accordance with the general instructions in section 7.2.1 of this manual.
- Clean the mounting plate with cleaning solvent.
- Inspect the whirling mechanism for wear. Repair or replace worn parts.
- Tighten all screws.
- Oil the shaft and other whirling parts.
- Replace the thermometers, making sure that they are mounted securely on the plate.
- Install a "fresh" wick.

Periodic maintenance requirements are:

- Change the wick every 2 weeks if the instrument is used daily. If used irregularly, change the wick when dirty or discolored.
- Inspect the whirling mechanism for wear.
- Store an uncased sling psychrometer out of the sunlight and in a clean location.

7.3.2 Wet- and Dry-Bulb Thermometers

Like a psychrometer, wet- and dry-bulb thermometers operate on the principle of cooling by evaporation. Ventilation is accomplished by the natural movement of the surrounding air, however.

a. Instrument

Wet- and dry-bulb thermometers are two spherical bulb thermometers, mounted side-by-side on a common backing, as illustrated in Figure 7.4. The wet bulb is covered by a wick which extends continually into a water container. Relative humidity and dew point temperature determinations obtained with this instrument generally are not as accurate as those obtained from a psychrometer because:

- The thermometers often do not have a full scale accuracy of better than 1 degree C.
- The spherical bulbs on the thermometers tend to be sluggish in response to atmospheric changes.
- The airflow across the thermometers may be inadequate, requiring artificial ventilation.

b. Installation

Wet- and dry-bulb thermometers are designed for installation in a shelter at a weather station, so that the thermometers are exposed to free circulation of air. The instrument, therefore, should be located in a position that allows for the natural movement of the surrounding air.

c. Operating instructions

Wet- and dry-bulb thermometers are read by the observer whenever a relative humidity and dew point temperature determination is desired. It is assumed that adequate ventilation of the wet bulb has been accomplished naturally.

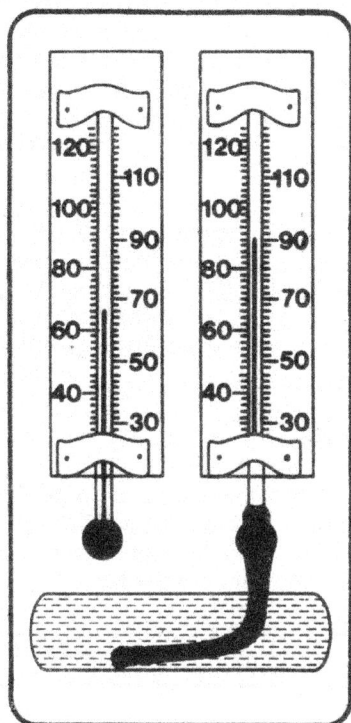

Figure 7.4. Wet- and dry-thermometers. The glass reservoir on the bottom must be kept full of water (Fischer and Hardy 1972).

To obtain the maximum level of accuracy possible, the observer should fan the thermometers, using a piece of cardboard, for about 3 minutes before taking the readings.

d. Maintenance

Wet- and dry-bulb thermometers are maintained in accordance with the general instructions in section 7.2.1 of this manual. An important item in the maintenance of this instrument is assuring that the thermometers agree within one-half of a graduation, when both thermometers are read as a dry bulb.

When replacing a broken thermometer, be sure that the replacement matches exactly the accuracy of the unbroken thermometer. To ensure this accuracy, it often is necessary to replace both of the thermometers with a matching pair. When replacing the thermometers, the wet bulb always hangs below the dry bulb, which minimizes the chance of moist air moving across the dry bulb during ventilation.

7.3.3 Hygrometer

There are several types of hygrometers, which are instruments that measure and record the relative humidity on a continuous basis. The simplest and the one most commonly used is the hair-element hygrometer. When a hair-element hygrometer is coupled with a temperature recording element, this instrument is called a "hygrothermograph," which is described in detail in the following section.

7.3.4 Hygrothermograph

A hygrothermograph simultaneously and continuously measures and records the air temperature and relative humidity (Figure 7.5). Although the details of construction differ by manufacturer, the general "working parts" of most hygrothermographs are similar.

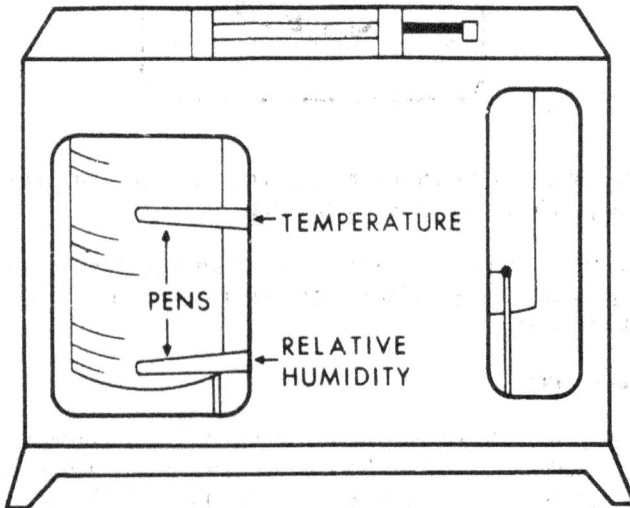

Figure 7.5. A hygrothermograph. The upper pen records temperature; the lower ppen records relative humidity (Fischer and Hardy 1972).

150

a. Instrument

The working parts of a hygrothermograph are:

- A temperature element.
- A relative humidity element.
- The pen arm assemblies.
- The chart drive mechanism.

The chart drive mechanism is a clock, either spring wound or battery driven, which turns a chart on a drum. The clock is located either inside the drum and turns with it or is fixed to the base of the hygrothermograph and the drum revolves around it. The pen arm assemblies are the link between the chart and the temperature and relative humidity sensors.

Most of the hygrothermographs used today contain deformation thermometers to measure the air temperature. Either a curved Bourdon tube or a bimetal strip that is curved or coiled forms the temperature element of these hygrothermographs.

The Bourdon tube is filled with an organic liquid. As the air temperature varies, the liquid in the tube expands or contracts, causing the Bourdon tube to stretch out or curl up, accordingly. These changes are transmitted to the chart through the pen arm linkage system in the instrument.

The bimetal strip is formed by welding together two different metals that have dissimilar expansion rates. As the air temperature changes, the metals expand or contract at different rates, causing the strip to change in form. These changes in form are transmitted to the chart through the pen arm system.

Most hygrothermographs use a human hair element to measure the relative humidity. This element is in the form of either a "bundle" of hairs or a "banjo spread" arrangement that resembles an open-faced bracelet with clasps on each end. The operating principle of these hygrothermographs is the same. High relative humidity results in a lengthening of the hair, while low relative humidity causes the hair to shorten. The movements of the hairs are transmitted to the chart through the pen arm linkage system.

b. Installation

A hygrothermograph usually is set in the left-front portion of a shelter. It is important that the instrument is placed far enough forward so it does not interfere with the whirling of a maximum-minimum thermometer.

c. Operating instructions

To remove an old chart:

- Lift the pens off the chart, using the shifting lever.
- Unlock and raise the cover.
- Lift the drum from the spindle, being careful not to damage the pens.
- Record the ending date and the "off time" on the chart.
- Pull the clips and remove the chart from the drum.
- Wind the clock.

To install a new chart:

- Write the station name or number, the beginning date, and the "on time" on the chart.
- Place the chart against the bottom lip of the drum and wrap the chart tightly around the drum.
- Insert the clip, if the cylinder is non-slotted. If the cylinder is lotted, insert the ends of the chart into the slot on the cylinder.
- Replace the drum on the spindle. Position the drum so that the chart time is a little faster than the actual time.
- If necessary, the pens should be refilled with ink.
- Return the pens to the chart, using the shifting lever. Check the ink flow by rotating the drum back and forth.
- Turn the drum to place the pens at the actual time on the chart by rotating the drum counterclockwise.
- Lower and latch the cover.

The calibration of the hygrothermograph should be checked whenever the instrument is visited. Because of the lag time of the sensors, the calibration checks should be made only when the air temperature and relative humidity are holding steady. The general procedure of a calibration check includes the following steps:

- Inspect the instrument for mechanical deficiencies.
- Remove any dust or dirt with a soft hair brush.

152

- To make a time mark, deflect each pen downward to obtain a short vertical line on the chart. Record the actual time near this mark. Compare the chart time with the actual time, and compare the time indicated with the upper pen with that of the lower pen. Make the necessary adjustments.
- Compare the maximum, minimum, and current temperatures recorded on the chart with the values obtained from a maximum, minimum, and dry-bulb thermometers. Make the necessary adjustments.
- Compare the current relative humidity recorded on the chart with a value obtained from a sling psychrometer. Observe the relative humidity "trace" on the chart for evidence of a range elongation or shortening. Make the necessary adjustments.

d. Maintenance

Because a hygrothermograph is a delicate instrument, the reliability of the data obtained from the instrument depends largely upon the level of maintenance. Due to the inherent characteristics of the sensors, relatively large errors occasionally can occur despite the best efforts of the observer. These errors can be minimized, however, through a continuous general maintenance program and calibration checks.

Carefully service the hygrothermograph and re-calibrate, as necessary, prior to each period of use, after changing the relative humidity element, and whenever a loss of calibration occurs. It is important to refer to the manual provided by the manufacturer of the instrument for the maintenance details. General maintenance items that apply to most hygrothermographs include:

- Keep the instrument clean. Remove dust and loose dirt with a soft brush. Brush lightly with a cleaning solvent to remove hardened dirt. Usually, only the clock needs to be oiled. Do not use cleaning solvent or oil on the hairs. Replace the hairs every 2 years.
- If the pens are clogged with dried ink, remove and wash in warm soapy water. If the pens fail to ink properly after cleaning, replace them. After proper inking is ensured, determine if both pens indicate the same time on the chart. If not, adjust one of the pens by changing its position on the pen arm. Make sure that the pens are attached firmly, so that they will not slide out of position in use.
- Remove all dirt and rust from the spindle and gears of the chart drive assembly. Lubrication of the spindle often is recommended to prevent rust. Make sure that the spindle is straight and that the

spindle and gears are fastened securely. Check for uniform backlash of an acceptable level, following the procedures specified in the manufacturer's manual. If the clock drive gear meshes too tightly with the large stationary gear at the bottom of the spindle, correct in accordance with the specific procedures in the manufacturer's manual.

- Check the clock by comparing the rate of drum revolution with the chart time scale. If necessary, use the regulator provided by the manufacturer to adjust the clock time. It is a good practice to have the clock cleaned and, if necessary, repaired by a competent repairman annually.

7.3.5   Data Collection and Processing

Atmospheric moisture data can be obtained with a psychrometer, wet- and dry-bulb thermometers, or from a hygrograph. With these data, the relative humidity and dew point temperatures can be calculated.

a.   Data collection

Relative humidity at a weather station often is determined on a daily basis. Measurements made with a psychrometer or wet- and dry-bulb thermometers, for example, can be used as the basis for this determination. Relative humidity is measured directly and recorded on a continuous basis with a hygrometer or a hygrothermograph.

b.   Data processing

Relative humidity determinations from wet- and dry-bulb temperatures, which is the case with the use of a psychrometer or wet- and dry-bulb thermometers, require the availability of "psychrometric tables," which are found in standard references on climatology and meteorology. Separate psychrometric tables are provided for different levels of atmospheric pressure and elevation above sea level.

Another type of "relative humidity table" presents values of relative humidity for different values of wet- and dry-bulb temperatures, with the latter values often referred to as the "wet-bulb depressions."

The collection and processing of data taken from a hygrometer or a hygrothermograph include chart annotations, tabulation of the data, and calculations. The chart annotations require the observer to:

154

- Enter the location of the instrument and the name of the observer.
- Enter the chart number and the dates of record.
- Enter the dates and watch time of chart placement, inspections, and removal.
- Note the "midnight" lines to denote clearly the extent of each day and to facilitate the reading of the graph.
- Enter notes to show any malfunctioning of the instrument or clock.

Tabulation of the data to calculate the mean daily relative humidity requires only one task, namely:

- Enter the daily maximum and minimum relative humidity values. These values generally are taken to represent the 24-hour time period before 8 AM local time.

The mean daily relative humidity value is calculated from:

$$RH_{mean} = \frac{RH_{max} + RH_{min}}{2} \tag{7.2}$$

where RH_{mean} = mean daily relative humidity (%)
RH_{max} = daily maximum relative humidity (%)
RH_{min} = daily minimum relative humidity (%)

The measurement of the "mean daily relative humidity" obtained through a solution of equation (7.2) is in reality a measure of the "median daily relative humidity."

7.4 EVAPORATION

The process of evaporation is the net loss of water from a surface by means of a change in the state of water from a liquid to a vapor, and the transfer of this vapor to the atmosphere. Evaporation from soils, plant surfaces, and water bodies, together with the water losses through plant leaves, a process referred to as "transpiration," are considered collectively in watershed management as "evapotranspiration."

The amount of evapotranspiration largely determines the proportion of the rainfall input to a watershed that becomes streamflow. The term "evapotranspiration" is synonymous, in a general sense, with "consumptive use," a term that refers to the amount of water that is required to mature a crop. When a vegetative cover is losing water to the atmosphere at a rate that is unlimited by any deficiencies in the water supply, the process is

155

known as "potential evapotranspiration." If the water supply is limited at some time in the year, the actual rate of evapotranspiration can be less than the potential rate.

The evaporative rate of the atmosphere usually is measured by one of two types of *evaporimeters*, an open pan or an atmometer. A number of methods have been used to measure the transpiration of plants, with the *closed-phytometer* generally the most favor by ecologists and watershed management specialists. Several methods are available to measure evapotranspiration. Some of these methods are applicable only to the determination of potential evapotranspiration rates, while others measure actual evapotranspiration. A measurement of evapotranspiration, either potential or actual, can be obtained through the use of a lysimeter (see section 7.4.4 of this manual).

7.4.1 Evaporation Pan

Because of the ease with which measurements can be made with an evaporation pan, this instrument has been used widely in watershed management. A pan exposing a free water surface has the feature of quantifying the evaporation rate in the same units as rainfall. However, the volume of water in the pan, the height of the rim above the water surface, the radius of the pan, and the color of the pan greatly influence the measurements obtained. All evaporation pans, therefore, should be constructed, installed, and operated in an indentical manner.

a. Instrument

The rate of evaporation from a small pan is not the same as that from a large body of water. Neither is the rate of evaporation the same for all types of pans or pan locations. Therefore, for a specific pan location, it is necessary to measure the rate of evaporation under all conditions and to determine the "pan coefficient" to be used in adjusting the measurements to those anticipated for a large body of water.

There are a number of commonly used evaporation pans and, therefore, it is difficult to describe a particular pan to use in a specific location. After a study of the results obtained from each of several types of evaporation pans, for example, the American Society of Civil Engineers recommended the U.S. Weather Bureau Class A land pan as a first choice (Figure 7.6). This pan is 1.219 m (4 ft) in diameter, 25.4 cm (10 in) deep, and the bottom is raised 15.24 cm (6 in) above the ground surface. The water surface is supposed to be at least 5.08 cm (2 in) and never more than 7.62

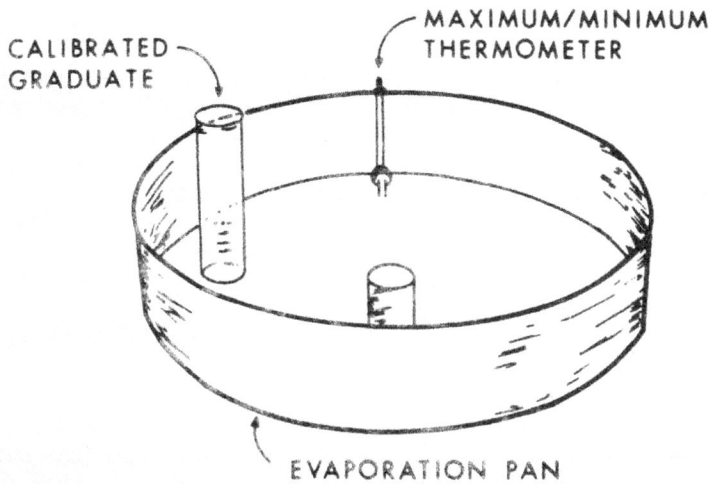

CALIBRATED GRADUATE

MAXIMUM/MINIMUM THERMOMETER

EVAPORATION PAN

Figure 7.6. A U.S. weather bureau class A evaporation pan.

cm (3 in) below the top of the pan. Other types of evaporation pans also are available and can be used, as appropriate.

b. Installation

The site for the installation of an evaporation pan, regardless of the type, should be level. Obstructions should not be closer to the pan than 4 times the height of the obstruction, and vegetative growth should not be higher than the height of the pan. The site should be selected to keep the water surface free of shadows.

The evaporation pan itself can be constructed of galvanized iron. Seams of the pan should be fabricated carefully to prevent the buckling of the bottom. A support platform for the pan can be made of lumber. The evaporation pan is centered on the platform, making sure that the bottom of the pan is level.

A stilling well for a hook gage, consisting of a hook at the end of a graduated stem, provides an undisturbed water surface for the measurement of evaporation from the pan.

157

c. Operating instructions

In obtaining measurements of evaporation in the pan, either a staff gage that is mounted on the side of the pan can be read, or a hook gage can be used to obtain more accurate measurements. A continuous water-level recorder also can be used to measure evaporation. Changes in the water level in the evaporation pan indicate the amount of evaporation.

Rainfall events that occur in the time interval between measurements must be taken into account in determining the rate of evaporation in the interval.

d. Maintenance

Vegetation surrounding an evaporation pan must not be allowed to grow above the level of the pan. The site on which the evaporation pan is installed should be fenced to prevent animals from drinking water from the pan.

The water level in an evaporation pan should be maintained at about 5 cm below the rim. A fluctuation of 2.5 cm from this level generally is allowable.

When a hook gage is used, it must be kept clean and oiled. Light machine oil applied 2 or 3 times a year can suffice. Before oiling, the hook gage should be cleaned thoroughly with a petroleum-based solvent.

7.4.2 Atmometer

The measurement of evaporation with an atmometer is a technique widely used by plant ecologists. Since no two atmometers lose water at the same rate, however, each instrument must be calibrated individually.

a. Instrument

An atmometer can be described in general terms as a thin sphere of unglazed, white pottery that is filled with water which soaks through the walls of the sphere and then evaporates into the air. By means of glass tubing and rubber stoppers, the sphere is connected to a reservoir bottle containing water, as shown in Figure 7.7. The amount of water lost by evaporation in a specified time interval is determined by measuring the amount of water that is necessary to refill the bottle up to a specified mark or by weighing the apparatus at the end of the time interval.

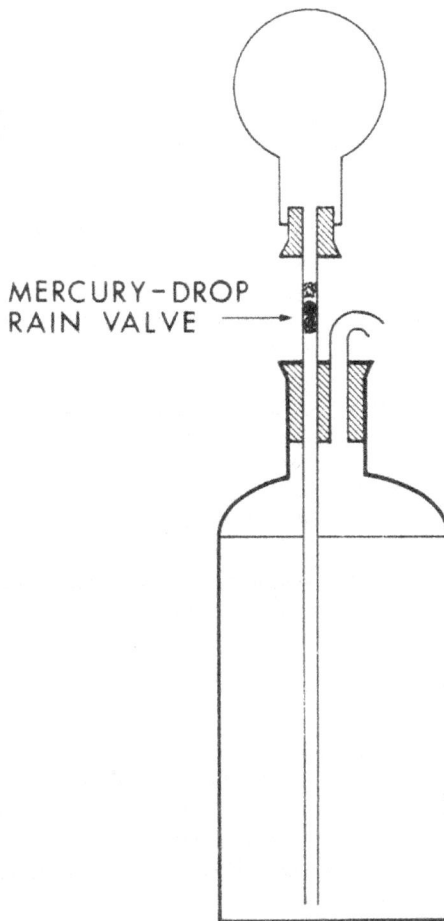

MERCURY-DROP RAIN VALVE ⟶

Figure 7.7. Diagram of an atmometer mounted on a reservoir bottle and equipped with a mercury-drop rain valve (Daubenmire 1956).

Evaporation easily pulls water from the bottle against gravity. However, when the outside of the sphere is wetted by rainfall, gravity would pull the water back into the apparatus. In this instance, the previous evaporation record would be lost, if it were not for a 5 to 8 mm layer of mercury installed in the delivery tube. An upward pull draws the water around the mercury, but a downward pressure flattens the mercury and stops the flow of water. The mercury layer, therefore, acts as a valve.

b. Operating instructions

Evaporation can be determined for any arbitrarily-specified time period by measuring the amount of water that is necessary to refill the reservoir bottle or weighing the atmometer.

For determining the evaporation rates in short time intervals, an atmometer can be connected to a burette, which permits readings to the nearest 0.01 ml. An appropriate length of glass tubing and rubber connections must be used to keep the porous sphere at a height above the water level in the burette. Otherwise, gravity will force the water into the sphere and increase the rate of water loss.

To calibrate an atmometer, the instrument, along with a calibrated one, is placed on a round table which is turned slowly by a breeze from an electric fan. In doing so, both atmometers are exposed to the same evaporative stress, and by comparing their water losses over a time period, a coefficient can be calculated for the instrument being calibrated. This coefficient then is applied to the collected data to adjust them to the standard. Since the coefficient can change slowly while the instrument is in use, a coefficient should be determined both before and after the period of calibration and then averaged.

c. Maintenance

When an atmometer becomes discolored or covered with algae, the instrument must be renovated by drying, rubbing with fine-grained sandpaper until the surface is once more white, and then scrubbing with a brush and water to remove the foreign particles from the pores. Under no circumstances should chemicals be added to the water as a deterent to the formation of algae.

7.4.3 Phytometer

The *closed-phytometer method* of measuring the transpiration rates of a plant often is employed by watershed managers. The instrument used in this method consists of a water-tight tank containing sufficient soil to nourish the plant. A cover is sealed onto the tank to prevent rain from entering and water from escaping from the tank, except through the process of transpiration by the plant. A method of adding water to the tank as desired is provided.

The transpirational losses for a specified time period are equal to the original weight of the phytometer plus the weight of the water added, minus the final weight of the instrument. A limitation to the use of the closed-phytometer method is the size of the water-tight tank. Therefore, the method generally is restricted to plants with relatively small root systems.

7.4.4 Lysimeter

A direct measurement of actual evapotranspiration can be obtained by analyzing the water balance for a block of soil. The soil is contained in a porous-bottomed tank, called a *lysimeter*, that is buried in the ground. The tank itself should be large and deep enough to reduce the boundary effects and to avoid the restriction of plant growth.

a. Instrument

A lysimeter can vary in capacity from 1 to over 150 m^3. Importantly, the soil profile, root development, and moisture conditions inside the lysimeter should be identical to those outside the lysimeter.

If the soil is maintained in a "wet" condition, evapotranspiration will be measured at the potential rates. If the purpose is to measure actual evapotranspiration, the moisture content of the soil in the tank should be allowed to fluctuate in a manner that is similar to the soil outside of the tank.

b. Operating instructions

A lysimeter generally is classified as either a weighing type or drainage type (Figure 7.8). A weighing type of lysimeter facilitates the evaluation of each element in the water balance equation:

$$\text{evapotranspiration} = \text{precipitation} + \text{irrigation} - \text{drainage} + \text{changes in storage} \qquad (7.3)$$

Changes in storage are measured by weighing the block of soil, which can be accomplished manometrically or with a scale system. Large weighing lysimeters are relatively expensive instruments that are used largely for research purposes.

In a drainage type of lysimeter, the water balance of the soil is assumed to be:

$$evapotranspiration = precipitation + irrigation - drainage \quad (7.4)$$

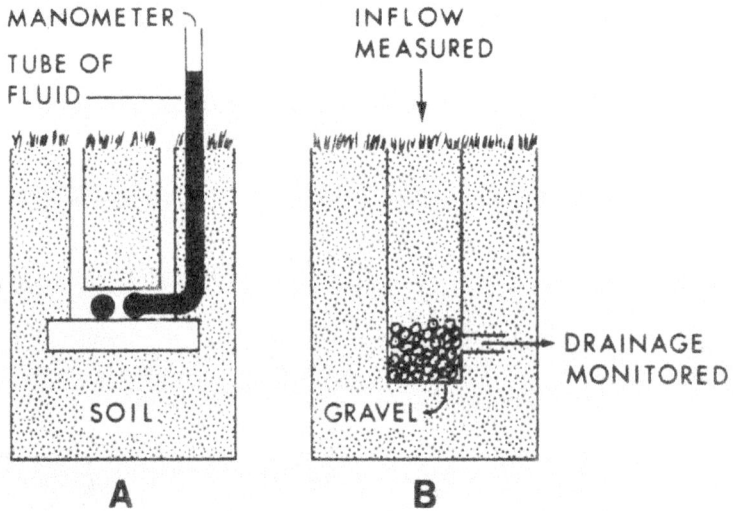

Figure 7.8. Schematic diagrams of a (A) weighing type and a (B) drainage type lysimeter (Dunne and Leopold 1978).

Water inflows and drainage are measured, but the changes in storage within the block of soil are not measured. The drainage type of lysimeter, therefore, is useful only in periods when the changes in storage are considered negligible. The length and timing of these periods depend upon the frequency of wetting, the size of the lysimeter, and the rate of movement of water through the soil.

7.4.5 Data Collection and Processing

Measurements of the evaporation for a 24-hour or longer period integrate the climatic factors that affect evaporation into the atmosphere, and also the hourly, daily, or periodic fluctuations in these factors. Measurements of evaporation from an open pan can be obtained in the same units as rainfall.

An atmometer can be used to measure the rate of evaporation, although the results obtained cannot be translated readily into the depth of water evaporated from a water surface. An atmometer is designed to generally simulate the surfaces of plants.

Measurements of the transpiration losses from plants that are obtained from a phytometer in a laboratory and measurements of evapotranspiration obtained from a lysimeter for research purposes are taken, in most instances, to represent a specified period of time.

a. Data collection

Open pan measurements frequently are taken so that daily, monthly and yearly evaporative rates can be determined. To measure the rate of evaporation on a daily basis, either a hook gage or a continuous water-level recorder should be used. Open pan measurements of evaporation frequently are taken at 8 AM local time.

b. Data processing

Measurements of evaporation from an open pan can be taken in units that are directly comparable to rainfall measurements. A pan coefficient for the installation site must be known to compensate for the differences in radiation, air temperature, wind, and heat storage between the pan and a larger body of water. Intervening rainfall amounts must be subtracted from the differences in measurements of evaporation from a pan.

A general procedure for the processing of evaporative data obtained on a continuous basis, including chart annotations, tabulation of the data, and calculations, must be specified. Chart annotations generally are:

- Enter the location of the instrument and the name of the observer.
- Enter the chart number and the dates of record.
- Enter the dates and watch times of chart placement, inspections, and removal.
- Note the "8 AM" lines to denote clearly the extent of each date to facilitate the reading of the graph.

Tabulation of the data involves one step:

- Enter the daily totals of evaporation, identifying the 24-hour time period used.

163

Calculations are:

- Daily totals of evaporation are calculated, from which monthly totals are obtained by summation.

7.5 WIND

Wind is air in motion. The motion of wind has two basic components: *wind velocity* and *wind direction*. Wind velocity refers to the rate at which air passes a point, while wind direction is simply the direction from which the wind is blowing.

7.5.1 Anemometer

Measurements of wind velocity are obtained from a cup anemometer that is exposed at a specified height above the ground. Regardles of how it is specified, the standard height must be adjusted to compensate for the height of the vegetative cover, surface irregularities, and nearby obstructions.

The World Meteorological Organization standard height for the placement of a cup anemometer is 10 m above the ground for wind measurement systems that supply general information on wind velocity for a watershed. For purposes of estimating potential evaporation, a cup anemometer should be installed at 2 m above the ground for collecting the required wind velocity data.

a. Instrument

Cup anemometers are calibrated to rotate at a rate that is proportional to the actual wind velocity. This rotation is transferred by a main shaft to either a contacting mechanism or a generator. Either the number of contacts made or the voltage generated is read on an indicating device that is wired to the anemometer. The readout device generally is located in a shelter or nearby building.

A contacting anemometer consists of a 3- or 4-cup rotor assembly, as shown in Figure 7.9, a main vertical shaft, a gear mechanism, and an electrical contact. In addition, some contacting anemometers contain a built-in dial or counter, which records and accumulates the total wind movement.

Figure 7.9. A 3-Cup Contact Anemometer (Fischer and Hardy 1972).

A generator anemometer has a rotor or cup assembly, a vertical shaft, a generator, and a device to indicate the wind velocity. The shaft connects the cups to a small permanent magnet generator. As the cups rotate, voltage is generated in proportion to the windspeed. Instantaneous indications of wind velocity are obtained from an attached voltmeter. It usually is difficult to obtain a mean wind velocity value from a generator anemometer.

Wind velocity also can be obtained by a hand-held anemometer. One example of such an instrument is the venturi-action meter, which is designed for quickly estimating the wind velocity near the ground.

b. Installation

An anemometer can be mounted on either a wooden pole or a iron pipe pole. Metal towers also can be used, especially in situations when anemometer heights in excess of 8 to 10 m. Metal towers are available in one-piece sections of a specified height, stacked sections that are extended and bolted together, and telescoping sections that crank up and down.

Regardless of how the anemometer is mounted, the installation should accomplish the following:

- Be wind-firm.
- Allow easy access to the anemometer.

- Accommodate the attachment of the readout device.
- Allow for periodic adjustment of the anemometer height.
- Be compatible with any existing lightning protection system.

c. Operating instructions

The specific operating instructions for an anemometer vary with the type of instrument. The manual provided by the manufacturer, therefore, should be consulted for these details. However, to illustrate the detail to which these operating instructions must be followed, those operating instructions specified for a contacting anemometer in which the readout is obtained by a reset counter equipped with a time are:

- Reset the dial to zero, if not already at zero.
- Set the time.
- When the timer stops, read the dial.
- Record the average 10-minute windspeed.
- Reset the dial to zero.
- If the average windspeed for other than 10 minutes is desired, set the desired time interval on the time and divide the final count bythat time interval.

The general operating instructions for a contacting anemometer in which the readout is obtained by a reset counter alone are:

- Reset the dial to zero, if not already at zero.
- Using the "off-on" switch, start both the counter and a stopwatch. If using a watch with a "sweep" second hand, start the counter as the hand passes 12.
- After 10 minutes, stop the counter.
- Record the average 10-minute windspeed.
- Reset the counter to zero.
- If the average windspeed for other than 10 minutes is desired, let the counter run for the desired time interval and divide the final count by that time interval.

The operating instructions of a contacting anemometer in which the readout is obtained by a non-reset counter are:

- Record the numbers on the counter dial.
- Start the counter and a stopwatch. If using a watch with a "sweep" second hand, start the counter as the second hand passes 12.
- After 10 minutes, stop the counter.

166

- Record the numbers on the dial.
- Calculate the average 10-minute windspeed by substracting the beginning count from the ending count.
- If the average windspeed for a period other than 10 minutes is desired, let the counter run for the desired time interval, substract the ending count from the beginning count, and divide the difference by the time interval used.

Operating instructions for a contacting anemometer in which the readout is obtained by a buzzer or flasher are:

- Close the switch on the buzzer or flasher.
- Immediately after the first flash or buzzer, record the time or start a stopwatch.
- Count the number of buzzes or flashes for the desired time interval.
- Open the switch.
- Calculate the average windspeed by dividing the total count by the time interval.

Operating instructions for a contacting anemometer with a self-contained readout counter are:

- Record the starting count.
- Record the count at the end of the time interval for which the average windspeed is desired.
- Substract the beginning count from the ending count and divide the difference by the time interval.

The operating instructions for a contacting anemometer with a self-contained readout dial are:

- Read the dial at the beginning of the time period for which the average windspeed is desired.

  - Read the inner dial first. The index for the inner dial is located in the outer dial. It is a small "zero" through which is drawn a vertical line.
  - Read the outer dial. The index is a small pointer located above and just to the left of the large dial. When the glass cover is on the dial, the observer's eye has to be lowered to see this second index.
  - The total reading is obtained by adding the indicated value on the outer dial to the value indicated on the inner dial.

Some features of this model include:

• Digital compass stores data with wind speed
• Automatic Crosswind calculator
• Headwind/Tailwind speeds
• Current, Maximum and Average Wind Speeds
• Air, Water & Snow Temperature
• Wind Chill
• Relative Humidity
• Dewpoint
• Heat Stress Index
• Barometric Pressure
• Altitude
• Density Altitude
• Wet Bulb Temperature
• Graph and recall trends
• Easy to read, backlit display
• Time and date
• Customize screens
• Flip-top impeller cover
• Automatically stores measurements
• Manually stores measurements
• Charts up to 1,400 measurements

This is a modern professional anemometer. For more details about this model, and a comparison with other models, see the following URL:

http://www.anemometers.co.uk/anemometer_kestrel_4500_anemometers.html

168

Many mechanical problems can be identified by spinning the cups and observing their action. Difficult starting and abrupt stopping can indicate a need for oil, a need for cleaning, bent or worn parts, or improper assembly. In normal operation, the cups should not wobble while spinning. Wobbling often indicates a bent shaft.

A major item to consider when inspecting an anemometer for mechanical soundness is the flow of electrical current from the anemometer to the readout device.

7.5.2 Wind Vane

The instrument used to measure wind direction is a wind vane.

a. Instrument

A wind vane is an asymmetrically-shaped pointer or arrow mounted on a vertical shaft, as illustrated in Figure 7.10. Importantly, the pointer or arrow must turn freely in "very light" winds. A wind vane may or may not be equipped with a readout device.

b. Installation

The two most important items to consider in the installation of a wind vane are:

- Proper orientation of the wind vane in relation to true north.
- Careful wiring of the readout device (if one is used) to the wind vane, so that the direction indicated on the readout device corresponds to the direction indicated by the wind vane.

c. Maintenance

A wind vane is designed for trouble-free operation for long periods of time. Annual maintenance consisting of cleaning, lubricating, and general refurbishing usually is sufficient to keep a wind vane in good operating condition.

7.5.3. Data Collection and Processing

The rate of the movement of air frequently is expressed in terms of the mean velocity of the wind in a specified time interval, such as an hour, a day, or a month. However, since the flow of air is not as constant as the

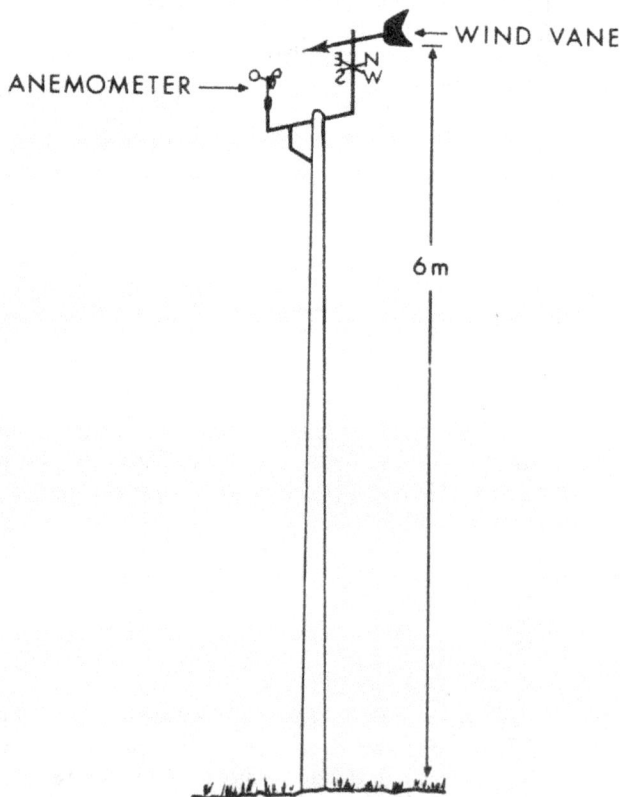

Figure 7.10. A wind vane mounted on a vertical shaft (Fischer and Hardy 1972).

flow of water in a stream, mean velocities of wind for long time intervals can be misleading. A wind of "gale proportions," for example, can blow for only a few minutes, but it may not be suspected that this event occurred from the magnitude of the hourly velocity.

In addition to the mean velocity of the wind, the total wind movement in a time period often is calculated to further describe the movement of air at a site.

a. Data collection

The data necessary to calculate the mean velocity of wind is collected in terms of the specified time interval. These data also can be used in obtaining the values of total wind movement for the specified time interval. In some instances, a measure of the *windspeed* at a site also is desired. Windspeed refers to the mean velocity of wind in a period of 10 minutes. In the operating instructions of an anemometer, therefore, a 10-minute record usually is specified. An instantaneous estimate of the wind speed can be obtained with a hand-held anemometer for any specified time period.

Wind direction usually is measured in terms of the duration of the flow of air from each of the "cardinal directions." The accuracy of these measurements depends upon the observer's knowledge of the cardinal directions in reference to a specific site.

b. Data processing

Wind movement is read manually or recorded continuously and automatically. In either case, the calculation of the mean wind velocity for a specified time period or the windspeed at a site is a straightforward procedure. For example, the mean wind velocity for a specified day, expressed in km/hr, is calculated by dividing the total wind movement (km) for the day by 24 hours.

When the data are recorded continuously on charts, the appropriate annotations, tabulations, and calculations must be specified. Chart annotations to obtain this measure of wind movement, for example, require the observer to:

- Enter the location of the instrument and the name of the observer.
- Enter the chart number and the dates of record.
- Enter the dates and watch times of chart placement, inspections, and removal.
- Note the "midnight" lines to denote clearly the extent of the calendar day, whenever the wind movement totals on this basis are desired.
- Wind velocity data often are correlated with 24-hour values of evaporation or other observed meteorological factors ending at 8 AM local time. In this case, mark the "8 AM" lines to facilitate the reading of the daily totals.

Tabulation of the data involves:

- Enter the daily wind movement totals, based on either of the 24-hour time periods, on an appropriate form. Identify the 24-hour time period used

Calculations include:

- Daily totals of wind movement, expressed in terms of km of wind movement, are calculated initially, from which monthly and yearly totals are obtained by summations.

7.6 SOLAR RADIATION AND RADIANT ENERGY

Solar energy and the radiant energy balance are critically important to life on earth. It is surprising, therefore, that only a few instruments were available for the measurement of solar radiation until recently. However, with increased interests in energy-budget relationships, a number of instruments have been developed and have become available commercially. These instruments generally measure different portions of the solar spectrum, have different optical and geometrical characteristics, and employ different methods of recording the data.

7.6.1 Units of Measurement

The *gram-calories* unit of measurement, when applied to solar radiation, sums the effects of invisible radiation and light. When measuring a small quantity of energy, the *joule* or *erg* often is used as the unit. The relationships among these units of measurements are:

$$1 \text{ gram-calorie (gm-cal)} = 4.18 \text{ joules (J)} = 41.8 \text{ million ergs}$$

Measurement of the intensity of light alone is based upon the illumination produced by a *standard candle*. The amount of light received at a distance of 1 m from a standard candle is called a *lux* or *meter-candle*. The lux generally has been accepted as the standard international unit of measurement for light intensity.

7.6.2 Pyranometer

The term *radiometer* refers generally to any instrument that measures radiation, regardless of the wavelength. However, in meteorology, the term implies an instrument that measures wavelengths from 0.3 to 100

172

microns and, therefore, includes the wavelengths that are important in transferring energy from the sun and earth. Such an instrument is called a *net (all-wave) radiometer* if its flat receptor is sensitive to both hemispheres and, as a result, measures the *radiant flux density* between them. The radiant flux density is defined as the amount of energy received on a unit surface in a unit of time. A *total (all-wave) hemispherical radiometer* is equipped with a flat receptor that is open to only a hemisphere.

Some radiometers only measure a restricted spectral range. A *pyrheliometer*, for example, measures the intensity of the direct solar beam radiation. This instrument, which has a relatively narrow field of view, is oriented so that the flat plate of the receiver is normal to the direct rays of the sun for measurement.

The instrument most commonly used to measure the total incoming solar radiation, which includes both direct solar beam and diffuse solar radiation, is a *pyranometer*, as described below. The spectral range of a pyranometer is restricted to short-wave radiation, from approximately 0.3 to 3 microns. The term *solarimeter* is interchangeable with pyranometer.

a. Instrument

One example of an electrical pyranometer is the Eppley pyranometer, although other types of electrical and mechanical pyranometers are available. The sensing element in an Eppley pyranometer is a thermopile with black and white segments. Since black absorbs and white reflects solar radiation, these segments develop different temperatures. The greater the radiant flux density, the greater is the temperature difference between the black and white areas. The temperature difference is sensed by a differential thermopile, whose output is nearly linear with the radiant flux density.

The glass cover for a pyranometer, regardless of the type, should be made of a material that permits the transmission of a major portion of the solar spectrum.

Important characteristics to consider in choosing a pyranometer include the following:

- The sensitivity should be able to detect physically or biologically significant differences in the radiant flux density.
- The temperature dependence should be known and correctable by electronic or computational means. Circuits to compensate for

changing temperatures can be incorporated into the instrument.
- The response time should be appropriate to the needed measurement of solar radiation. If the sensors are too slow, the measurements can be incorrect. If the sensors are too fast, the record can contain more detail than is required.
- An instrument that senses solar radiation or light must respond appropriately, according to the cosine of the angle of incidence of the impinging solar radiation stream. Deviations from the cosine can result in measurement errors.

There are other types and styles of electronic pyranometers available commercially. In addition, mechanical pyranometers also have been used widely. Some of these mechanical pyranometers work on the same principle as the Eppley pyranometer, with the temperature differences registered by non-electrical means. Other pyranometers are based on somewhat different principles. For example, with the Bellani pyranometer, the absorbed solar radiation is converted into heat that vaporizes alcohol. The quantity of alcohol that is condensed at the base of the instrument is proportional to the solar radiation received.

To measure the diffuse solar radiation alone, a pyranometer must be shaded from the direct beam solar radiation and, therefore, exposed only to the sky radiation. This shading can be accompished by using a shadow band that is designed to cast a narrow shadow over the sensing disk of the pyranometer. The shadow band should be adjustable to accommodate the changing solar elevation and changing azimuth of sunrise and sunset as the year advances. These latter changes are relatively minor in the ASEAN region and, for the most part, can be ignored. The output of a shaded pyranometer also must be adjusted upward to account for the scattered solar radiation from that portion of the sky that the shadow band blocks from view.

With the total solar radiation and the diffuse solar radiation measured, the direct beam solar radiation fraction is calculated as:

$$R_s(\text{direct beam}) = R_s(\text{total}) - R_s(\text{diffuse}) \qquad (7.5)$$

b. Installation, operating instructions, maintenance

The installation of a pyranometer should follow the procedures specified in the manual provided by the manufacturer. In all cases, however, care must be taken to level and securely fasten the instrument. When an electrical pyranometer is selected for use, special care should be taken to ensure "good" electrical contacts in the lead wires.

Measurements of solar radiation obtained with a pyranometer can be taken either continuously or discretely, depending upon the type and style of instrument and the objectives of the measurements. Continuous measurements can be obtained on a chart placed on a drum that is activated by a timing mechanism. Discrete measurements require an observer to simply record the solar radiation readings at specified time intervals.

The glass cover for the pyranometer should be cleaned regularly. The instrument should be covered with a protective cloth when not in use. As wet surfaces will introduce measurement errors, the pyranometer should be dried carefully after rainfall events.

7.6.3 Net Radiometer

A net radiometer absorbs radiation of all wavelengths directed downward (toward the surface) and upward (away from it). The absorbing surfaces of this instrument, therefore, should behave as "black body" absorbers for all wavelengths.

a. Instrument

The sensing element of a net radiometer is a differential thermopile that is separated by an insulating material, so that each blackened absorbing surface developed a temperature that is proportional to the radiant flux density impinging on it. The difference in temperature is translated into a difference in voltage output of the thermopile.

A net radiometer generally is shielded with a plastic dome that is transparent to both the visible and infrared wavelengths. An inflated plastic dome of polyethylene or polystyrene covers the sensor, protecting it from rainfall. The transparency of the plastic is not identical for long-wave and short-wave radiation. Some net radiometers are not covered with plastic domes, but instead they are ventilated by a stream of air on both sides of the sensing plate.

Similar to a pyranometer, the performance of a net radiometer is judged in terms of its sensivity, temperature dependence, response time, and cosine response. However, the calibration of a net radiometer usually is less stable than that of a pyranometer, due to the aging of the blackened sensing surfaces and the plastic dome.

b. Installation, operating instructions, maintenance

The same general comments made on the installation, operating instructions, and maintenance of a pyranometer also apply to a net radiometer. Frequent checks against a "standard" instrument and field or laboratory calibrations are desirable.

7.6.4   Total Hemispherical Radiometer

A net radiometer of either the shielded or unshielded type can be adapted to measure the total radiation, both short-wave and long-wave, on one side or the other of the sensor. Such an instrument then becomes a total hemispherical radiometer.

a. Instrument

To measure the total hemispherical radiation from the sky, for example, the bottom surface of a shielded net radiometer is covered with a metallic dome that is blackened on the inside. A thermocouple measures the internal surface temperature of the metal dome. The long-wave radiation impinging on the lower surface of the sensor is calculated from:

$$E = k(T^4) \tag{7.6}$$

$$\text{where } E = \text{radiant energy flux of a black body (gm-cal/min)}$$
$$k = \text{Stefan-Boltzmann constant (ly/min)}$$
$$T = \text{absolute temperature of the black body (degrees K)}$$

The Stefan-Boltzmann constant is calculated for a specified set of conditions by multiplying $(8.132 \times 10^{-11} \text{ ly}) \times (\text{degrees K}^{-4}) \times (\text{min}^{-1})$.

The total (all-wave) hemispherical radiation is the summation of the measured net radiation and the long-wave radiation calculated by equation (7.5).

b. Installation, operating instruction, maintenance

The installation, operating instructions, and maintenance considerations for a pyranometer and a net radiometer also are appropriate for a total hemispherical radiometer.

7.6.5 Sunshine Duration

It is possible to estimate the amount of solar radiation from the data on the sunshine duration or cloudiness at a site. For example, one type of "sunshine switch" transmits a signal when the light intensity is sufficiently strong to activate an electrical circuit.

A mechanical sunshine duration meter, the Campbell-Stokes sunshine recorder, consists of a glass globe that acts as a magnifying glass, focusing a beam of sunlight onto a special recording paper when the sunlight is sufficiently strong. A trace is burned on the recording paper as the sun moves through the sky. This trace indicates the duration of "bright" sunlight. The depth of the burn also is an indicator of the intensity of the solar radiation.

7.6.6 Data Collection and Processing

Radiant flux densities generally vary with the time of day, the time of the year, and the current meteorological conditions. It would appear necessary, therefore, to take continuous measurements, especially if a relatively detailed radiation balance climatology is to be developed for a site. However, the specific measurement requirements of a monitoring program usually will permit a reduction in the number of measurements required.

a. Data collection

In taking measurements of radiant flux densities, one initially must determine the data requirements. These measurements frequently are taken to represent a daily mean value, a value at noon, a seasonal total, or some other arbitrary value. To determine a daily mean value or a seasonal total, a continuous record usually is necessary.

Measurements of solar radiation or sunlight above a vegetative canopy are relatively easy to make and analyze. However, measurements of radiant flux densities under a vegetative canopy are more difficult to obtain, especially when complicated patterns of light and shadow are present. Little generally is known about this type of variation or about suitable methods of sampling in these conditions.

b. Data processing

In some cases, the number of observations needed to obtain a representative data set can be reduced through a stratification of the days

177

by weather-type or season. To do so, however, requires the development of a weather-typing scheme that is suitable to the area of concern.

When continuous measurements have been obtained, chart annotations, tabulation of the data, and calculations are specified. The chart annotations required generally include:

- Enter the location of the instrument and the name of the observer.
- Enter the chart number and the dates of record.
- Enter the dates and watch times of chart placement, inspections, and removal.
- Note either the "midnight" or "8 AM" lines, depending upon the basis of the data summarization, to facilitate the reading of the graph.

Tabulation of the data requires:

- Enter either the daily totals of the radiant flux density on the appropriate form.

Calculations include:

- Daily total values are summed to obtain monthly values and then the monthly values are summed to obtain the yearly value.

As stated earlier, whether continuous measurements, periodic instantaneous readings, or periodic integrated measurements should be taken is dictated largely by the objectives of the monitoring program. In many instances, continuous measurements are converted to periodic values by reading time-sequence values from the chart. In these cases, little information is lost if the radiation values were sampled periodically in the first place. It can be more efficient, therefore, to use a point-printing or digitizing recorder to scan a number of measuring instruments in a time sequence or to manually scan and read a number of instruments.

7.7   DATA ANALYSIS

7.7.1   Statistical Analysis

Data sets representing long-term, high-quality measurements of rainfall, air temperature, relative humidity, evporation, wind, and solar radiation and radiant energy generally are summarized and analyzed statistically. Two of the commonly used methods of statistical analysis are regression and frequency analysis.

a. Regression analysis

Regression analysis requires the selection of appropriate mathematical models that describe the relationships between a dependent variable (for example, air temperature) and one (time of day) or more (time of day and elevation) independent variables. These mathematical models are approximations which are based on sample data and, therefore, are subject to sampling variations. Unfortunately, in many instances, these sampling variations are unknown.

A *simple regression*, which defines a relationship between one dependent (Y) and one independent variable (X), is used with a straight-line relationship between the two variables. Other simple regressions describe non-linear relationships, for example, parabolic, exponential, or logarithmic relationships.

The general mathematical model for a simple regression is:

$$Y = a \pm b(X) \qquad\qquad (7.7)$$

where Y = dependent variable
  X = independent variable
  a, b = regression coefficients estimated by statistical formulas, which are found in references of statistical analysis.

A dependent variable frequently is related to more than one independent variable, in which case a *multiple regression* is defined. If a relationship can be estimated by a multiple regression, it allows for a more precise estimate of the dependent variable than is possible by a simple regression.

The general mathematical model for a multiple regression is:

$$Y = a \ b(X_1) \ c(X_2) \dots \qquad\qquad (7.8)$$

b. Frequency analysis

A frequency analysis of data sets is performed for many purposes, for example, determining the probability of a rainfall event of a specific magnitude (as described in section 2.5.3 of this manual) or delineating a watershed in terms of the probabilities of windspeeds of particular magnitudes occurring. A definite pattern of frequency occurrence of units in each of a series of equal classes is a "frequency distribution function."

For a data set that represents a population of interest, a frequency distribution function shows the relative frequencies that different values of a variable (X) will occur. By knowing the frequency distribution function, it is possible to determine what proportions of the individuals in the population are within specific "size" limits. Each set of data representing a population has its own frequency distribution function. There are certain general types of functions that often occur in watershed management, however, including the *normal*, the *binomial*, and the *Poison* distributions.

The normal frequency distribution, which is a "bell-shaped" distribution, is used frequently in the statistical analysis of watershed measurements. Theoretically, a normal frequency distribution, which is illustrated in Figure 7.11, exhibits the following properties:

- The mean, median, and mode are indentical in value.

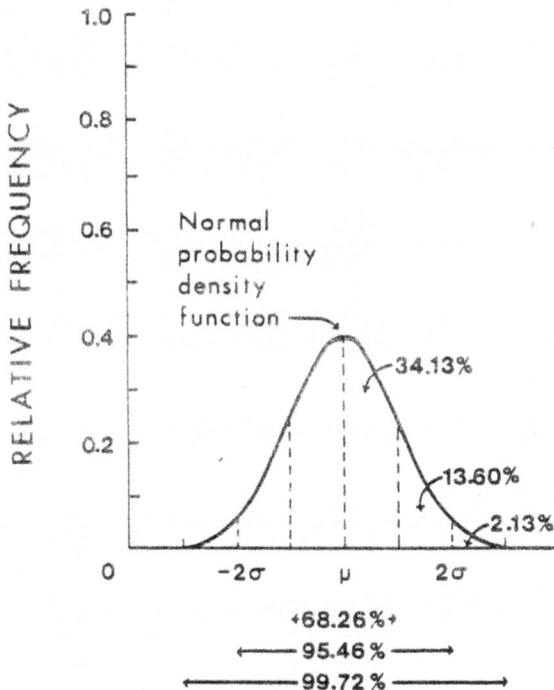

Figure 7.11. A normal frequency distribution (Brooks et al. 1990).

- Small variations from the mean occur more frequently than large variations from the mean.
- Positive and negative variations about the mean occur with equal frequencies.

A binomial frequency distribution is associated with data where a fixed number of the individuals are observed on each unit. Furthermore, the unit is characterized by the number of individuals having some specific attribute.

An asymmetric Poisson frequency distribution can arise where individual units are characterized by a count having no fixed upper limit, particularly if zero or low counts dominate.

There are a number of other frequency distribution functions that can be used to describe populations of watershed attributes. One of the more important tasks of a statistican is to identify the most appropriate function for a situation, and then apply the "proper" statistical methods to estimate the population parameters and, when necessary, to test hypotheses about these parameters.

7.7.2 Time Series Analysis

Climatic data sets frequently occur in the form of a time series, resulting in questions of the following kind often arising: "For a specific climatic measurement (for example, air temperature), is there evidence that a change in the series of measurement is occurring, and if so, what is the nature and magnitude of the change?" Many other questions of this kind often come to the attention of watershed managers and planners.

Available statistical procedures, such as student's "t" test for estimating and testing a change in the mean values, can play a role in the analysis of changes in climatic data sets through time. However, a student's "t" test generally is valid only if the observations in the data set are independent and vary about the mean values normally with constant variance. A climatic data set is often in the form of a time series in which successive observations are serially dependent, non-stationary, and with a "strong" seasonal effect. Commonly-employed parametric and non-parameteric statistical procedures, which rely on independence or special symmetry in the distribution function, therefore, are not always appropriate in the analysis of changes in climatic measurements through time. An alternative approach is a time series analysis.

It is beyond the scope of this manual to describe the theory and technical application of time series analysis techniques in detail. This background information is available in references on time series analysis and stochastic processes. Here, however, it is important to mention that in applying a time series analysis, or in general any type of stochastic analysis, to climatic data sets, three steps in analytical development are recommended:

- The first step is to identify the form of the "mathematical model" that "fits" the climatic data sets in question.
- In the second step, which is the estimation stage, the model parameters are calculated by using the method of maximum likelihood of occurrence.
- Finally, the model is "checked" for possible inadequacies. If these diagnostic checks reveal a serious anomaly, the appropriate model modifications are made by repeating the first two steps.

There have been a number of theoretical and technical advancements in time series analysis techniques recently. These advancements, coupled with the availability of computer facilities, allow these techniques to be used with relative ease, assuming that long-term, high-quality data collection have been made.

References

American Public Health Association. 1976. Standard Methods for the Examination of Water and Wastewater. American Public Health Association, Washington, D.C., USA.

American Society of Civil Engineers. 1969. Design and Construction of Sanitary and Storm Sewers. Manual and Reports on Engineering Practices 37, American Society of Civil Engineers, New York, USA.

American Society for Testing and Materials. 1976. Standard for Metric Practice. American Society for Testing and Materials, Washington, D.C., USA.

Anderson, H. A., M. D. Hoover, and K. G. Reinhart. 1976. Forests and Water: Effects of Forest Management on Floods, Sedimentation, and Water Supply. General Technical Report PSW-18, Southwest Forest and Range Experiment Station, U.S. Department of Agriculture, Berkeley, California, USA.

ASEAN-US Watershed Project. 1985. Proceedings of the Symposium on Watershed Management and Conservation for Productive and Protective Uplands in the ASEAN Region. ASEAN-US Watershed Project, College, Laguna, Philippines.

ASEAN-US Watershed Project. 1986. Proceedings on the Workshop on Standardization of Guidelines for Watershed Management Approaches and Researches in the ASEAN Region. ASEAN-US Watershed Project, College, Laguna, Philippines.

Avery, T. E. 1975. Natural Resources Measurements. McGraw-Hill Book Co., New York, USA.

Beard, L. R. 1962. Statistical Methods in Hydrology. U.S. Army Corps of Engineers, Sacramento, California, USA.

Blake, G. J. 1975. The Interception Process. In: Prediction in Catchment Hydrology. National Symposium on Hydrology, Australian Academy of Science.

Box, G. E. P., and G. M. Jenkins. 1970. Time Series Analysis: Forecasting and Control. Holden-Day, Inc., San Francisco, California, USA.

Brakensiek, D. C., H. B. Osborn, and W. J. Rawls. 1979. Field Manual for Research in Agricultural Hydrology. Agriculture Handbook 224, Soil and Water Conservation Research Division, Agricultural Research Service, U.S. Department of Agriculture, Washington, D.C., USA.

Branson, F. A., G. F. Gifford, K. G. Renard, and R. F. Hadley. 1981. Rangeland Hydrology. Kendall/Hunt Publishing Co., Dubuque, Iowa, USA.

Brooks, K. N., P. F. Ffolliott, H. M. Gregersen, and J. L. Thames. 1990. Hydrology and the Management of Watersheds. Iowa State University Press, Ames, Iowa, USA. (In preparation)

Brown, G. W. 1980. Forestry and Water Quality. Oregon State University Bookstores, Corvallis, Oregon, USA.

Brown, H. E. 1969. A Combined Control-Metering Section for Gaging Large Streams. Water Resources Research 5:888-894.

Brown, H. E., E. A. Hansen, and N. E. Champagne, Jr. 1970. A System for Measuring Total Sediment Yield from Small Watersheds. Water Resources Research 6:818-826.

Brown, R. M., N. I. McClelland, R. A. Deininger, and M. F. O'Connor. 1973. Water Quality Index - Crashing the Psychological Barrier. Advances in Water Pollution Research. Pergamon Press, Oxford, England, pp. 878-897.

184

Buchanan, T. J., and W. P. Somers. 1969. Discharge Measurements at Gaging Stations. In: Techniques of Water-Resources Investigations of the United States Geological Survey. Geological Survey, U.S. Department of the Interior, Reston, Virgina, USA.

Bureau of Reclamation. 1975. Water Measurement Manual. Bureau of Reclamation, U.S. Department of the Interior, Denver, Colorado, USA.

Cochran, W.. G. 1963. Sampling Techniques. John Wiley & Sons, Inc., New York, USA.

Chunko, K., N. Tang Tham, and S. Ungkulpakdikul. 1971. Measurements of Rainfall in the Early Wet Season Under Hilt and Dry-Evergreen, Natural Teak, and Dry-Dipterocarp Forests in Thailand. Forest Resources Bulletin 10, Faculty of Forestry, Kasetsart University, Bangkok, Thailand.

CODEL. 1980. Simple Assessment Techniques for Soil and Water. Environment and Development Program, CODEL, New York, USA.

Colman; E. A. 1953. Vegetation and Watershed Management: An Appraisal ofVegetation Management in Relation to Water Supply, Flood Control, and Soil Erosion. The Ronald Press Co., New York, USA.

Daubenmire, R. F. 1956. Plants and Environment: A Textbook of Plant Autocology. John Wiley & Sons, Inc., New York, USA.

Dunne, T., and L. B. Leopold. 1978. Water in Environmental Planning. W. H. Freeman and Co., San Francisco, California, USA.

Federer, C. A. 1970. Measuring Forest Evapotranspiration. Research Paper NE-165, Northeast Forest Experiment Station, U.S. Department of Agriculture, Upper Darby, Pennsylvania, USA.

Fischer, W. C., and C. E. Hardy. 1972. Fire-Weather Observers' Handbook. Intermountain Forest and Range Experiment Station, U.S. Department of Agriculture, Ogden, Utah, USA.

Freitchen, L. J., and L. W. Gay. 1979. Environmental Instrumentation. Springer-Verglag, New York, USA.

Geological Survey. 1977. National Handbook of Recommended Methods for Water-Data Acquisition. Office of Water Data Coordination, Geological Survey, U.S. Department of the Interior, Reston, Virginia, USA.

Gray, D. M. 1970. Handbook on the Principles of Hydrology. Water Information Center, Inc., Port Washington, New York, USA.

Green, R. H. 1979. Sampling Designs and Statistical Methods for Environmental Biologists. John Wiley & Sons, Inc., New York, USA.

Hansen, E. A. 1966. Field Test of an Automatic Suspended Sediment Sampler. Transactions of the American Society of Agricultural Engineers 9:739-743.

Hewlett, J. D. 1982. Principles of Forest Hydrology. University of Georgia Press, Athens, Georgia, USA.

Holtan, H. N., N. E. Minshall, and L. L. Harrold. 1962. Field Manual for Research in Agricultural Hydrology. Agriculture Handbook 224, Soil and Water Conservation Research Division, Agriculture Research Service, U.S. Department of Agriculture, Washington, D.C., USA.

Horowitz, J. L. 1969. An Easily Constructed Shadow-Band for Separating Direct and Diffuse Solar Radiation. Solar Energy 12:543-545.

Houghton, D. D. 1985. Handbook of Applied Meteorology. John Wiley & Sons, Inc., New York, USA.

Inhaber, H. 1976. Environmental Indices. John Wiley & Sons, Inc., New York, USA.

Jones, K. R. 1981. Arid Zone Hydrology for Agricultural Development. FAO Irrigation and Drainage Paper 37, Food and Agriculture Organization of the United Nations, Rome, Italy.

King, H. W. 1954. Handbook of Hydraulics. McGraw-Hill Book Co., New York, USA.

Kittredge, J. 1948. Forest Influences. McGraw-Hill Book Co., New York, USA.

Kunkle, S. H., and J. L. Thames. 1977. Guidelines for Watershed Management. FAO Conservation Guide 1, Food and Agriculture Organization of the United Nations, Rome, Italy.

Kunkle, S. H., W. S. Johnson, and M. Flora. 1987. Monitoring Stream Water for Land-Use Impacts. Water Resources Division, National Park Service, U.S. Department of the Interior, Fort Collins, Colorado, USA.

Lee, R. 1978. Forest Microclimatology. Columbia University Press, New York, USA.

Lee, R. 1980. Forest Hydrology. Columbia University Press, New York, USA.

Leupold and Stevens, Inc. 1978. Stevens Water Resources Data Book. Leupold and Stevens, Beaverton, Oregon, USA.

Linsley, R. K., Jr., M. A. Kohler, and J. L. H. Paulhus. 1975. Hydrology for Engineers. McGraw-Hill Book Co., New York, USA.

Manokaran, N. 1979. Streamflow, Throughfall, and Rainfall Interception on a Low Tropical Rain Forest in Peninsular Malaysia. The Malayan Forester 42:174-201.

McCoy, J. W. 1969. Chemical Analysis of Industrial Water. Chemical Publishing Co., New York, USA.

Middleton, W. E. K., and A. F. Spilhaus. 1953. Meteorological Instruments. University of Toronto Press, Toronto, Canada.

Midgley, D., and K. Torrance. 1978. Potentiometric Water Analysis. John Wiley & Sons, Inc., New York, USA.

Northern Region Agricultural Development Center. 1981. Workshop Manual for the Analysis of Water Samples. Northern Region Agricultural Development Center, Chiang Mai, Thailand.

Ponce, S. L. 1980. Water Quality Monitoring Programs. WSDG Technical Paper 00002, Watershed Systems Development Group, Forest Service, U.S. Department of Agriculture, Fort Collins, Colorado, USA.

Raros, R.S. 1979. Critical Ecological Considerations in River Basin Management in Southeast Asia. Paper Presented at the Regional Training Course on Integrated River Basin Management, College, Laguna, Philippines.

Read, G. W. 1975. Water Test Kit I: User's Manual. Bureau of Water and Environmental Research, University of Oklahoma, Norman, Oklahoma, USA.

Reifsnyder, W. E., and H. W. Lull. 1965. Radiant Energy in Relation to Forests. Technical Bulletin 1344, U.S. Department of Agriculture, Washington, D.C., USA.

Roehl, J. W. 1962. Sediment Source Area Delivery Ratios and Influencing Morphological Factors. In: IASH Publication 59, Gentbrugge, Belgium.

Rosenberg, N. J., B. L. Blad, and S. B. Verma. 1983. Microclimate: The Biological Environment. John Wiley & Sons, Inc., New York, USA.

Rosgen, D., K. L. Knapp, and W. F. Megahan. 1980. Total Potential Sediment. In: An Approach to Water Resource Evaluation, Nonpoint Source Silviculture, EPA-600 18-80-012, Athens, Goergia, USA

Sanders, T. G., R. C. Ward, J. C. Loftis, T. D. Steele, D. D. Adrian, and V. Yevjevich. 1983. Design of Networks for Monitoring Water Quality. Water Resources Publications, Littleton, Colorado, USA.

Satterlund, D. R. 1972. Wildland Watershed Management. The Ronald Press, Co., New York, USA.

School of Renewable Natural Resources. 1988. Resource Development of Watershed Lands. International Training Course Syllabus, School of Renewable Natural Resources, College of Agriculture, University of Arizona, Tucson, Arizona, USA.

Seller, W. D. 1965. Physical Climatology. University of Chicago Press, Chicago, Illinois, USA.

Smith, L. P. 1970. The Difficult Art of Measurement. Agriculture Meteorology 7:281-283.

Steel, R. G. D., and J. H. Torrie. 1960. Principles and Procedures of Statistics. McGraw-Hill Book Co., New York, USA.

Swanson, R. H., and R. Lee. 1966. Measurement of Water From and Through Shrubs and Trees. Journal of Forestry 64:187-190.

Swenson, H. A., and H. L. Baldwin. 1965. A Primer on Water Quality. Geological Survey, U.S. Department of the Interior, Washington, D.C., USA.

Tangtham, N. 1981. Ecological Aspects of Water Resources Management in Humid Tropical Ecosystems. Paper Presented at the Regional Training Course on Watershed Management and Environmental Monitoring, Chiang Mai, Thailand.

Trewartha, G. T. 1954. Introduction to Climate. MaGraw-Hill Book Co., New York, USA.

U.S. Environmental Protection Agency. 1971. Methods for Chemical Analysis of Water and Wastes. U.S. Environmental Protection Agency, Washington, D.C., USA.

U.S. Environmental Protection Agency. 1976. Quality Criteria for Water. U.S. Environmental Protection Agency, Washington, D.C., USA.

U.S. Environmental Protection Agency. 1979. Handbook for Analytical Control in Water and Wastewater Laboratories. Environmental Monitoring and Support Laboratory. U.S. Environmental Protection Laboratory, Cincinnati, Ohio, USA.

U.S. Water Resources Council. 1976. Guidelines for Determining Flood Flow Frequency. Bulletin 17, U.S. Water Resources Council, Washington, D.C., USA.

Wells, W. G., II, and P. M. Wohllgenmuth. 1987. Sediment Traps for Measuring Onslope Surface Sediment Movement. Research Note PSW-393, Pacific Southwest Forest and Range Experiment Station, U.S. Department of Agriculture, Berkeley, California, USA.

William, E. J. 1959. Regression Analysis. John Wiley & Sons, Inc., New York, USA.

Wisler, C. O., and E. F. Brater. 1965. Hydrology. John Wiley & Sons, Inc., New York, USA.

World Meteorological Organization. 1969. Guide to Meteorological Instrument and Observing Practices. World Meteorological Organization, Geneva, Switzerland.

Zinke, P. J. 1967. Forest Interception Studies in the United States. In: Forest Hydrology. Pergamon Press, Oxford.

www.ingramcontent.com/pod-product-compliance
Lightning Source LLC
Chambersburg PA
CBHW021541260326
41914CB00001B/107